How is Digitalization Affecting Agri-food?

Using real cases of food firms and agriculture supply chains as a context, *How is Digitalization Affecting Agri-food? New Business Models, Strategies and Organizational Forms* aims to understand the key themes in strategic and organizational research in this area.

Despite the importance of food and agriculture in the current political and societal context, analysis of the impact of digitalization and information technologies on the industry is still limited. The objective of this monograph is to understand the direction of this change. With case studies of food firms and agriculture supply chains it sets out to conceptualize food organizing and organizations as a fruitful object of inquiry, both at the intra and interorganizational levels. It aims to understand new business models, strategies, and organizational forms. Contributions in this stream of research have the potential to yield important and relevant insights for both scholars and societies.

This book is written primarily for academics engaged in innovation management or strategy, or conducting organizational behavior research. It will also be of relevance to practitioners and managers in the agri-food industry.

Maria Carmela Annosi is Assistant Professor of Innovation Management and Organizational Behaviour at the School of Social Sciences, Wageningen University and Research, The Netherlands. She is co-founder of DigiMetis center of excellence on the analysis of socio-ecological systems in a digitalized era, inside Wageningen University and Research.

Federica Brunetta is currently Assistant Professor of Management and Competitive Strategy at the Department of Business and Management of Luiss Guido Carli University in Rome. She is the Vice Director of the Research Centre for Business Transformation (ReBooT) at the LUISS Business School.

Routledge Studies in Innovation, Organizations and Technology

Society and Technology
Opportunities and Challenges
Edited by Ewa Lechman and Magdalena Popowska

Contextual Innovation Management
Adapting Innovation Processes to Different Situations
Patrick van der Duin and Roland Ortt

Research, Innovation and Entrepreneurship in Saudi Arabia
Vision 2030
Edited by Muhammad Khurram Khan and Muhammad Babar Khan

Developing Digital Governance
South Korea as a Global Digital Government Leader
Choong-sik Chung

Digital Business Models
Perspectives on Monetisation
Adam Jabłoński and Marek Jabłoński

Developing Capacity for Innovation in Complex Systems
Strategy, Organisation and Leadership
Christer Vindeløv-Lidzélius

How is Digitalization Affecting Agri-food?
New Business Models, Strategies and Organizational Forms
Maria Carmela Annosi and Federica Brunetta

For more information about this series, please visit: www.routledge.com/Routledge-Studies-in-Innovation-Organizations-and-Technology/book-series/RIOT

How is Digitalization Affecting Agri-food?
New Business Models, Strategies and Organizational Forms

Written and edited by Maria Carmela Annosi and Federica Brunetta

LONDON AND NEW YORK

First published 2021
by Routledge
2 Park Square, Milton Park, Abingdon, Oxon OX14 4RN

and by Routledge
52 Vanderbilt Avenue, New York, NY 10017

Routledge is an imprint of the Taylor & Francis Group, an informa business

© 2021 selection and editorial matter, Maria Carmela Annosi and Federica Brunetta; individual chapters, the contributors

The right of Maria Carmela Annosi and Federica Brunetta to be identified as the authors of the editorial material, and of the authors for their individual chapters, has been asserted in accordance with sections 77 and 78 of the Copyright, Designs and Patents Act 1988.

All rights reserved. No part of this book may be reprinted or reproduced or utilised in any form or by any electronic, mechanical, or other means, now known or hereafter invented, including photocopying and recording, or in any information storage or retrieval system, without permission in writing from the publishers.

Trademark notice: Product or corporate names may be trademarks or registered trademarks, and are used only for identification and explanation without intent to infringe.

British Library Cataloguing-in-Publication Data
A catalogue record for this book is available from the British Library

Library of Congress Cataloging-in-Publication Data
Names: Annosi, Maria Carmela, editor. | Brunetta, Federica, editor.
Title: How is digitalization affecting agri-food?: new business models, strategies and organizational forms / edited by Maria Carmela Annosi and Federica Brunetta.
Description: First Edition. | New York : Routledge, 2021. | Series: Routledge studies in innovation, organizations and technology | Includes bibliographical references and index.
Identifiers: LCCN 2020024068 (print) | LCCN 2020024069 (ebook)
Subjects: LCSH: Food industry and trade–Technological innovations. | Food supply. | Management–Technological innovations. | Organizational change.
Classification: LCC HD9000.5 .H689 2021 (print) | LCC HD9000.5 (ebook) | DDC 664.0068/4–dc23
LC record available at https://lccn.loc.gov/2020024068
LC ebook record available at https://lccn.loc.gov/2020024069

ISBN: 978-0-367-19651-6 (hbk)
ISBN: 978-0-429-20370-1 (ebk)

Typeset in Bembo
by Newgen Publishing UK

Contents

List of contributors ix
Foreword xi

Introduction 1
MARIA CARMELA ANNOSI AND FEDERICA BRUNETTA

Presentation of the chapters 3

PART I
Theories and concepts 9

1 **The impact of digitalization: a review on the issue** 11
   MARIA CARMELA ANNOSI, FEDERICA BRUNETTA, AND
   LI PEICHAN

   Introduction 11
   Systematic literature review: the methodology 12
   Results 15
   Conclusion 20

2 **Digital technology in the agri-food sector: a review on the business impact of digitalization in the agri-food sector** 25
   MARIA CARMELA ANNOSI AND FEDERICA BRUNETTA

   Introduction 25
   Digital technologies in agri-food 26
   Methodology 27
   Discussion 28
   Conclusion 32

PART II
Challenges and current strategies in digitalization in agri-food 35

3 **The role of managers or owners of SMEs in driving the digitalization process in the agri-food sector** 37
   IVAN BEDETTI, MARIA CARMELA ANNOSI, GIORGIA BUCCI, DEBORAH BENTIVOGLIO, WILFRED DOLFSMA, AND ADELE FINCO

   Introduction 37
   Methods 38
   Findings 39
   Conclusion 46

4 **The use of digital technologies in agri-food: Evidences from the rural sector** 49
   MARIA CARMELA ANNOSI AND FEDERICA BRUNETTA

   Synopsis 49
   Case A: the successful case of technology use in Company A 49
   Founder and manager of Company A 50
   Case B: a story of technology change, the case of Company B 54
   Founder and manager of Company B 54

PART III
The emerging promises of Internet of Things (IOT) and digital ecosystems in agri-food 59

5 **The power of business ecosystems in the agri-food sector: an interview with Sjaak Wolfert** 61
   MARIA CARMELA ANNOSI AND TEUN GILISSEN

   Background 61
   Interview 62

6 **Research insights on the governance and dynamics of IOT business ecosystems** 74
   TEUN GILISSEN, MARIA CARMELA ANNOSI, AND WILFRED DOLFSMA

   Introduction 74
   Business ecosystems 76

Emerging collective actions within business ecosystems: cooperation or coopetition 78
Governance mechanisms within business ecosystems 79
Contractual and relational governance 80
Forms of governance 81
Business ecosystem dynamics 82
Conclusion 88

7 The social impact of ICT-enabled interventions among rural Indian farmers as seen through eKutir's VeggieLite intervention   93
SPENCER MOORE, MARIA CARMELA ANNOSI, TEUN GILISSEN, JENNIFER MANDELBAUM, AND LAURETTE DUBE

Introduction 93
Social impact 94
VeggieLite intervention 94
Social brokerage among rural farmers 96
Conclusion 97

PART IV
The evolution of digitalization in agri-food   99

8 Artificial intelligence: toward a new economic paradigm in agri-food   101
ALESSIO GOSCIU, FEDERICA BRUNETTA, AND MARIA CARMELA ANNOSI

Introduction 101
Main applications of AI 102
How is AI supporting agri-food? 103
Conclusion 104

9 Agri-food and AI: integrating technology as a core element of the business model: the case of Elaisian   106
ALESSIO GOSCIU AND FEDERICA BRUNETTA

Introduction 106
Market analysis 107
Competition analysis 108
Technological readiness 109
Artificial intelligence and the Elaisian case 109

10 **Convergence research and innovation digital backbone: behavioral analytics, artificial intelligence, and digital technologies as bridges between biological, social, and agri-food systems** 111

LAURETTE DUBE, SJAAK WOLFERT, KARIN ZIMMERMAN, NATHAN YANG, FERNANDO DIAZ-LOPEZ, RIGAS ARVANITIS, SANDRA SCHILLO, SABINA HAMALOVA, JIAN YUN NIE, AND SHAWN BROWN

Introduction 111
Convergence research and innovation (CI) 113
Behavioral analytics 116
Person-centered systems 118
Multi-stakeholder decision support to agri-food innovation 120
Conclusion 122

Conclusion: Where next? 126

FEDERICA BRUNETTA AND MARIA CARMELA ANNOSI

Digitalization in agri-food: where next? 126
Avenues for research 127

Index 131

Contributors

Rigas Arvanitis, Institut de Recherche pour le Développement (IRD).

Ivan Bedetti, master's degree student, Wageningen University.

Deborah Bentivoglio, post-doc, department of agricultural, food and environmental sciences, Università Politecnica delle Marche.

Shawn Brown, director, Pittsburgh Supercomputing Center (PSC), a joint research center of Carnegie Mellon University and the University of Pittsburgh.

Giorgia Bucci, Ph.D. candidate, department of agricultural, food and environmental sciences, Università Politecnica delle Marche.

Fernando Diaz-Lopez, director, innovation for sustainable development network, associate professor extra-ordinary, department of industrial engineering, Stellenbosch University.

Wilfred Dolfsma, chair business management and organization group, Wageningen University.

Laurette Dube, professor, marketing; James McGill chair of consumer and lifestyle psychology and marketing; chair and scientific director, McGill Centre for the Convergence of Health and Economics.

Adele Finco, full professor, department of agricultural, food and environmental sciences, Università Politecnica delle Marche.

Teun Gilissen, research assistant, Wageningen University.

Alessio Gosciu, Luiss University Graduate.

Sabina Hamalova, research assistant, McGill University.

Jennifer Mandelbaum, Ph.D. candidate, department of health promotion, education, and behavior. Arnold School of Public Health.

Spencer Moore, associate professor, health promotion, education, and behavior, Arnold School of Public Health.

Jian Yun Nie, full professor, Universite de Montreal.

Li Peichan, master's degree student, Wageningen University.

Sandra Schillo, associate professor, Telfer school of management, University of Ottawa.

Sjaak Wolfert, senior scientist, Wageningen Research.

Nathan Yang, assistant professor, marketing, McGill University.

Karin Zimmerman, senior scientist and research leader, research infrastructures, Wageningen Research.

Foreword

It is with great pleasure that I write this foreword for the book *How is Digitalization Affecting Agri-food?* The agricultural sector is not only key to economies worldwide, it is the entry point by which nature and most traditional societies open to modernity.

In turn, food, before and beyond being an industrial activity, is core substance that mankind is made of. As the world enters its fourth industrial revolution that blurs the boundaries between biological, physical, social, and digital domains, digitalization may serve as a powerful balancing force in aligning activities in agri-food, industrial, and other social and economic sectors of traditional and modern society toward a sustainable future for generations to come. This requires deep integration of disciplines, knowledge, theories, methods, data, and communities from many fields of research and sectors of practice. Silos are to be eliminated to create systematic cohesion, thinking, and, most importantly, real-world transformation at scale.

This book, by management scholars Maria Carmela Annosi and Federica Brunetta, with contributors of diverse expertise, is an important move in this direction.

Laurette Dube, Ph.D., MBA
James McGill Chair, Desaultels Faculty of Management
Founding Chair and Scientific Director, McGill Center
for the Convergence of Health and Economics

Introduction

Maria Carmela Annosi and Federica Brunetta

The exponential growth of digital technologies is affecting various economic sectors and the way industries operate and perform. This digital transformation is also occurring in the agri-food sector, creating a new competitive pressure on incumbents to induce their senior managers or owners to include such technologies on their leadership agendas (Fitzgerald et al., 2014; Hess et al., 2016; Singh and Hess, 2017). It is introducing new economic rules and business models (BMs) with the involvement of new players that now move more easily across industries and sectors. Through these digital platforms, the fundamentals of supply and demand are also changed. Digital platforms and their service providers are now covering new crucial roles in orchestrating resources within the emerging platform-based business ecosystems and are reshaping how companies compete. In such ecosystems, firms have to upgrade their own managerial competencies as well as their organizational capabilities.

In this landscape, business ecosystems centered on firms' needs to exploit the opportunities derived from the adoption of new digital technologies are booming. Traditional approaches to leading businesses are therefore increasingly replaced by a tendency to build a strong, multi-layered network of partners that exchange knowledge and create a market for digital solutions for farming and food production.

Business ecosystems are introducing new pressure for firms' strategic renewal, including: (1) a renewed BM; (2) a new collaborative approach with new partners in the business ecosystem; (3) a new culture favoring the exchange of information and an openness to learning and adapting; and (4) the acquisition of new skills to step up the speed and scale of change together with the definition of new roles, such as a more diverse set of digital product owners.

However, there are few empirical research studies that analyze how organizations face digital transformation (Warner and Wäger, 2019), especially in the agri-food sector. Incumbents also experience important challenges when senior management strongly supports the digital transformation of processes, structures, and BMs (Hess et al., 2016; Warner and Wäger, 2019). One of the main problems firms might have is the lack of proper capabilities allowing them to overcome past path dependency (Svahn et al., 2017), to successfully discover and develop new opportunities, to integrate internally- and externally-generated new ideas, to protect internal intellectual property rights, to update "best practice" business processes, and to reshape the "rules of the game" in the global marketplace.

Although these issues are highly relevant, the development of capabilities for digital transformation is a pivotal issue. This book illustrates how small and medium-sized enterprises (SMEs) operating in the agri-food sector are organizing their business to face the difficulties of digital transformation.

In this book, we consider digital transformation as "the use of new digital technologies (social media, mobile, analytics or embedded devices) to enable major business improvements such as enhancing customer experience, streamlining operations, or creating new business models" (Fitzgerald et al., 2014:2). We also maintain that digital transformation is "an organizational transformation that integrates digital technologies and business processes in a digital economy" (Liu et al., 2011:1730). Indeed, embracing the perspective of Singh and Hess (2017:124), we also argue that the term "transformation," rather than "change," better underlines how the digital transformation within a firm needs to be considered as the "comprehensiveness of actions" that must be adopted to meet the new opportunities derived from the application of new digital technologies or reduce the possibility that new threats may impact the competitivity of firms. We also believe that "digital transformation is fundamentally not about technology, but about strategy," (Rogers, 2016: 308), therefore we mostly focus on how senior managers or owners of firms need to identify ways to exploit opportunity and design BM innovations.

In making sense of the previous literature on the challenges that digital transformation has created for incumbent firms in the agri-food sector, we address how managers of SMEs in this sector have perceived the change, strategized the usage of new technology, and organized their relevant activities. Additionally, drawing on senior managers' experiences with projects related to digitalization, we have initially discovered potential managerial factors affecting firms' reactions to digitalization waves and explaining the high incidence of failure regardless of the digitalization objective they may have.

To further explore these micro-foundations, we also present two teaching cases on the digital transformation of two Italian incumbents. For both cases, given the relevance of senior manager or owner in SMEs, we use the perspective of the senior leader (Regnér, 2008; Whittington, 2006), underlining the individual manager or owner's standpoint on practicing strategic activities for digital transformation. This investigation helps in further understanding the digital transformation in the agri-food sector by bringing participant perspectives (Gioia et al., 2013). Using the polar types approach, which includes the description of two opposite cases, we present a study of distinct patterns in data, relationships, and logics. Additionally, through a dedicated interview with the project manager of SmartAgriHubs and the EU project Internet of Food and Farm, we also uncover the underlining dynamics of ongoing business ecosystems. SmartAgriHubs is an on-going EU project devoted to boosting the digital transformation of the European agri-food sector. It aims to build a strong, multi-layered network of agricultural ecosystems providing firms with access to the latest technologies and expertise and allowing to test and experiment. The main purpose of this project is to activate and extend the current ecosystems established with the Internet of Food and Farm, which is approaching its end, by creating a network of hubs for digital innovation boosting the uptake of digital technologies and solutions.

The experiences recalled in the vibrant interview with Sjaak Wolfert will open a theoretical discussion on the issue of emerging governance mechanisms in the absence of any coordination in the business ecosystem. Two exemplary cases of business ecosystem in agri-food sectors will help readers understand the concepts and reflect on the implications they have on SMEs in agri-food sectors.

We conclude this book with the introduction of the next wave of digitalization in the agri-food sector; these include blockchain technology, automation, and artificial intelligence (AI) technology.

Presentation of the chapters

The chapters in this book are grouped into four sections, recalling the different interdependent perspectives underlined above:

- Theories, concepts, and definitions;
- Experienced challenges and current strategies organizations adopt to face digitalization in the agri-food sector;
- Dynamics of digital business ecosystems in the agri-food sectors;
- Evolution of digitalization in the agri-food sector's transformation.

Theories, concept and definitions

The section starts with an overview of the history of the most common applications of digital system in farming operations, retracing the steps accomplished by the industrial and agriculture revolution. As industry is progressing at a much faster rate than agriculture, we clarify how the development of digital technology is affecting the agri-food sector differently than the other sectors. Additionally, we present the peculiar challenges addressed by SMEs within the agri-food sector that allow for the practical implementation of Industry 4.0.

After defining waves of digitization in agriculture and the main progresses compared to other sectors, terms of digitization and digitalization are described. We also clarify how digitalization influences diverse business activities, including BMs, offerings, and relationships with other actors.

Together with this, we offer relevant highlights on the extant management research take into account the role of digitalization. The main purpose of this section is threefold: (a) to present a holistic view of the research, (b) to discuss the applicability of the main empirical results collected in SMEs enterprises operating in the agri-food sector, and (c) to provide a proposal for future research.

We then proceeded by performing an investigation of scholarly work linking digitalization and agri-food. We started by defining 4.0 technologies within the agri-food industries and then focused on the factors that might hamper the adoption and diffusion of such technologies in agri-food, by relying on extant literature that has empirically analyzed such issues.

Challenges and current strategies organizations adopt to face digitalization in the agri-food sector

In this section, we discuss the apparent contradiction that while digitalization is pervasive and impacts the organizational practices of any company, what is happening in the agri-food sector seems to be absent from the relevant research discourse in management literature. Thus, we present our further understanding about the phenomenon by relying on the early results collected from our research projects. We show how the organizational change derived from the adoption or integration of new digital technologies is mainly constrained and defined by the concurrent streams of social actions that managers or owners of SMEs can initiate or are involved in both internally and externally to the company. Our qualitative research emphasizes entrepreneurial activities that occur in SMEs, particularly those that compete in traditional and agricultural markets. In these organizations, technology does not constitute a relevant pillar for the creation and identification of new opportunities. Thus, unlike technology-based organizations, technological search is not usually conducted by well-trained staff members, instead conducted informally by agricultural entrepreneurs exploring the technological frontiers falling outside their traditional business areas. Acquisition of knowledge and information about new technology happens through the informal network of the managers' contacts, made up of peers, friends, and suppliers. Through regular interactions with all these players, managers can accumulate their knowledge, expanding their ability to absorb changes, and promote technological prospecting which constitutes a search for technologies both informal and formal. In line with this perspective, and through the presented teaching cases, we offer insights on how managers perceive, interpret, and develop strategies to formalize new technologies.

Furthermore, as illustrated by the teaching cases, the approach to the search process undertaken by managers can be very different, from passive discovery to being very active in determining the nature and definition of new business opportunities.

The dynamics of digital business ecosystems in the agri-food sectors

The role of inter-organizational networks and business ecosystems has gained traction over the years (Capaldo, 2007; Szeto, 2000; Teece, 2010). Studies not only indicate inter-organizational networking as a prominent feature of innovation activity in firms (Peters et al., 2010), but also point out how relationships and collaborations between actors in business ecosystems can lead to innovation activity (Davis, 2016; Dhanaraj and Parkhe, 2006).

In line with this perspective, social networks, and more in general, business ecosystems, are currently used in the agri-food sector as a means to trigger the partial diffusion of digital innovations throughout parts of the established networks involving SMEs. However, many problems have emerged since the first application of business ecosystems in the agri-food sector, including the involvement of low-skilled staff, and SMEs are suffering from scarcity of organizational resources for technological investment. Opportunities and challenges faced by current business ecosystems are discussed in the vibrant and engaging interview with Sjaak Wolfert, the project coordinator of Wageningen Digital Innovation hubs, created through the IoF EU project (www.iof2020.eu) and potentiated further through Wageningen SmartAgriHubs (http://smartagrihubs.eu). Both projects bring together stakeholders from various fields (ICT, finance, law, agriculture, research, governments) and involve all innovation players (start-ups, SMEs, large companies, farmers, food producers and retailers, regulatory bodies, and educational institutions). Through their recurrent experiments with innovation, there is greater potential for digital innovation hubs, triggering the dynamics of spreading knowledge within the entire ecosystem. Both projects monitor the ecosystem's dynamics and assess their impacts on farmers' attitude toward technology, on the organization of farmers' resources, and on practices of adopting technology.

However, from an academic point of view, the emerging questions regarding business ecosystems still remain; these include: (1) how to integrate diverse entities coherently? (2) how to allocate resources, activities, and property rights so that each firm, the community, and the ecosystem as a whole can gather benefits? and (3) How to collaborate and coordinate?

To date, there is still a scarcity of literature that provides a theoretical, analytical, and empirical grounding of the governance of digital business ecosystems, especially in the agri-food sector. This section presents different perspectives of the governance of digital business ecosystems and suggests more focused future research activities.

This section ends with a description of two exemplary cases that demonstrate in-depth the structure of social networks involving, among other entities, SMEs operating in agri-food sector. Through their presence in the business ecosystems, these SMEs, as potential adopters of innovations, can more easily access information about innovations that can allow them to adopt new technologies. In addition, the cases also describe the application of other models of expanding innovation, such as the effect of contagion, social influence, and social learning. Indeed, thanks to this network, people can come into contact with others who have already adopted new digital technologies (contagion). Moreover, through their belonging to the ecosystem, people can better understand what innovation is worth adopting, benefitting from the outcomes experienced by prior adopters (social learning). Additionally, within the network, there is a level of conformity that results from people within the ecosystem adopting innovations when they have evidence of others in the group presenting successful outcomes from their own previous experiences.

Evolution of digitalization in the agri-food sector's transformation

The agri-food sector is subject to diverse pressure, that are often highlighted throughout the book, but above all by those related to the need of reducing costs and/or creating value, not only for competitive reasons but also for the scrutiny related to sustainability challenges that require meeting the needs of a growing world population (FAO, 2013). It is evident that, in order to cope with these challenges, innovation will be key (Morand and Barzman, 2006).

In this latter section, we focus on the use of technology to meet the objectives of economic and environmental sustainability. We focus on the role of AI, behavioral analytics, and digital technologies as tools to achieve superior advantage while bridging between biological, social, and agri-food systems for integrated sustainable development. We present the case of Elaisian, a precision farming service, that thanks to a system of algorithms based on a database of agronomic studies and machine learning techniques has developed tools that are able to prevent diseases and to optimize cultivation processes, such as irrigation and fertilization.

These are important issues within the industry's technological evolution. Thus, any delay in adopting technology and innovation exposes companies not only to increased threats of competition but also to societal imperatives related to environmental challenges.

We conclude this introduction by thanking the colleagues who have supported our work with their precious contributions, and in particular to Laurette Dube, Sjaak Wolfert, Spencer Moore, Wilfred Dolfsma, Adele Finco, Karin Zimmerman, Joerg Nessing, Shawn Brown, Jennifer Mandelbaum, Nathan Yang, Fernando Diaz-Lopez, Rigas Arvanitis, Sandra Schillo, Sabina Hamalova, Jian Yun Nie, Deborah Bentivoglio, and Giorgia Bucci. We are also very grateful to our promising students Teun Gilissen, Ivan Bedetti, Alessio Gosciu, and Peichan Li for their noteworthy work and contributions.

References

Capaldo, A. (2007). Network structure and innovation: the leveraging of a dual network as a distinctive relational capability. *Strategic Management Journal*, 28(6), 585–608.

Davis, J. P. (2016). The group dynamics of interorganizational relationships: collaborating with multiple partners in innovation ecosystems. *Administrative Science Quarterly*, 61(4), 621–661.

Dhanaraj, C., & Parkhe, A. (2006). Orchestrating innovation networks. *Academy of Management Review*, 31(3), 659–669.

FAO (2013). Climate-smart Agriculture – Sourcebook in Nations. In: FAO (Ed.). *Food and Agriculture Organization of the United Nations*, Rome, Italy.

Fitzgerald, M., Kruschwitz, N., Bonnet, D., & Welch, M., (2014). Embracing digital technology: a new strategic imperative. *Sloan Management Review*, 55(2), 1–1.

Gioia, D. A., Corley, K. G., & Hamilton, A. L. (2013). Seeking qualitative rigor in inductive research: notes on the Gioia methodology. *Organizational Research Methods*, 16(1), 15–31.

Hess, T., Matt, C., Benlian, A., & Wiesboeck, F. (2016). Options for formulating a digital transformation strategy. *MIS Quarterly Executive*, 15(2), 123–139.

Liu, D. Y., Chen, S. W., & Chou, T. C. (2011). Resource fit in digital transformation: lessons learned from the CBC Bank global e-banking project. *Management Decision*, 49(10), 1728–1742.

Morand, F., & Barzman, M. (2006). European sustainable development policy (1972–2005): fostering a two-dimensional integration for more effective institutions. [University works] auto-saisine. 34 p. hal-01189947.

Peters, L. D., Pressey, A. D., Westerlund, M., & Rajala, R. (2010). Learning and innovation in inter-organizational network collaboration. *Journal of Business & Industrial Marketing*, 25(6), 435–442.

Regnér, P. (2008). Strategy-as-practice and dynamic capabilities: steps towards a dynamic view of strategy. *Human Relations*, 61(4), 565–588.

Rogers, D. L. (2016). *The digital transformation playbook: rethink your business for the digital age*. New York: Columbia University Press.

Singh, A., & Hess, T. (2017). How chief digital officers promote the digital transformation of their companies. *MIS Quarterly Executive*, 16(1), 1–17.

Svahn, F., Mathiassen, L., & Lindgren, R. (2017). Embracing digital innovation in incumbent firms: how Volvo cars managed competing concerns. *MIS Quarterly*, 41(1), 239–253.

Szeto, E. (2000). Innovation capacity: working towards a mechanism for improving innovation within an inter-organizational network. *The TQM Magazine*, 12(2), 149–158.

Teece, D. J. (2010). Business models, business strategy and innovation. *Long Range Planning*, 43(2–3), 172–194.

Warner, K. S. & Wäger, M. (2019). Building dynamic capabilities for digital transformation: an ongoing process of strategic renewal. *Long Range Planning*, 52(3), 326–349.

Whittington, R. (2006). Learning more from failure: practice and process. *Organization Studies*, 27(12), 1903–1906.

Part I
Theories and concepts

1 The impact of digitalization

A review on the issue

Maria Carmela Annosi, Federica Brunetta, and Li Peichan

Introduction

Assisted by digital technologies, digitalization has changed society enormously. Digitization is recognized as the digitalized formats and data that are able to change the business or business model (Gobble, 2018). It is assisted by digital technologies as well as the combinations of information, computing, communication, and connectivity (Bharadwaj et al., 2013). Digitalization is empowering, refining, and renovating current businesses as well as society. Digital technologies are indeed relevant from both an economic and social point of view, as they create disruptions, change the boundaries of entire industries, and change the rules of competition (Christensen, 2007; Porter and Heppleman, 2014).

Yoo (2012) proposes a more comprehensive definition of digitalization, describing it as "the encoding of analog information into a digital format and the subsequent reconfiguration of the socio-technical context of production and consumption of the product and services."

The resulting "4.0 Era," which represents the so-called "Fourth Industrial Revolution," is providing organizations with both opportunities and challenges, imposing profound transformations to organizations, politics, and society. Technologies underlying digitalization have changed not only the existing way companies are doing business but have also created brand-new companies. Together with digitalization, the impact is far-reaching from both a macro and micro point of view. They are part of the organizational core processes (Alaimo and Kallinikos, 2017), while at the same time transforming entire ecosystems in the way information is produced, controlled, and circulated (Scott, 1995).

Digitalization undoubtedly modifies the company's operational and commercial activities (Calia et al., 2007) as well as strategies and business models (Barrett and Davidson, 2008; Bharadwaj et al., 2013). At the same time, entire ecosystems are recreated (Teece, 2010).

In order to keep the business alive and retain competitive advantages, companies need to react to digitalization and digital technology innovation properly, often implementing changes at the business model level (Zott and Amit, 2010). Business models (BM) are commonly recognized as the way that organizations create, deliver, and capture values in various contexts, such as economic, social, and cultural (Osterwalder and Pigneur, 2010).

Immense literature about Business Models has emerged, as well as technology innovation. However, there is not much literature that focuses on how digitalization, with its digital technology as an innovation, affects the way companies are doing business. Therefore, this research will focus on how digitalization impacts business in the context of business model design and strategy. Literature reviews are an important part of any research project, and a systematic literature review can provide researchers with a precise reviewing process (Tranfield et al., 2003). In this chapter, we present a systematic literature review to better understand digitalization in business.

Systematic literature review: the methodology

This research followed a three-stage procedure, including planning the review, conducting the review, and reporting and dissemination (Tranfield et al., 2003).

In this first stage, the objective of the research and the key data source were identified, with a specific focus on the database Web of Science. The second stage of conducting the review contained five steps: (1) identification of research, (2) selection of studies, (3) study quality assessment, (4) data extraction and monitoring progress, and (5) data synthesis.

Instead of using digitalization as a keyword solely, words with wider meanings were selected, including "Digital*," "Cyber*," "Big data," "AI," "Artificial intelligence," "Industry 4.0," and "Smart*." Meanwhile, to avoid overlap, the operator "OR" was used to connect all keywords selected while searching. (TS=Digital* OR Cyber* OR Big data OR Artificial intelligence OR AI OR Industry 4.0 OR Smart*). Keywords were used as a selection criterion for the topic. We restricted the time frame of research to 2008–2018. Besides keywords and publication years, the initial search of database was set as: document types "article"; language "English"; and subject area "business" and "management." To further restrict the quality of the database, the sources of the articles in the database had to be clustered. In this research, only 4 and 4* journals according to Academic Journal Guide 2018 (ABS, 2018) were selected. As such, all the articles in the database are of high quality in business and management fields.

This initial search gained a total of 632 articles, which were then further analyzed. A synthesis of the outcomes from the top ten journals publishing related articles is shown in Table 1.1.

Grouping

To further classify the articles, we grouped them in distinct categories. The grouping method followed the research by Crossan and Apaydin (2010). The first group (Group 1) contains only reviews and meta-analysis. The second

Table 1.1 Top ten journals publishing digitalization research

Source title	Records	% of 632
Mis Quarterly	78	12.34
Information Systems Research	73	11.55
European Journal of Operational Research	67	10.60
Journal of Management Information Systems	56	8.86
Tourism Management	46	7.28
Management Science	45	7.12
Marketing Science	40	6.33
Journal of Product Innovation Management	30	4.75
Research Policy	28	4.43
Journal of Marketing	26	4.11

Table 1.2 Grouping result of initial dataset

Group	Initial dataset
Group 1 Reviews and meta-analyses	67
Group 2 Highly cited articles	419
Group 3 Recent articles	213
Total	**632**

group (Group 2) consisted of selected papers centered on citation-based selection criteria for the initial dataset. The last group (Group 3) included residual articles from the initial dataset. There were no duplicate articles within each group.

- **Group 1 Reviews and meta-analyses.** To select reviews and meta-analyses, "review" or "meta" were added in the search term based on topic (title, keywords, or abstract). (TS = "Digital* OR Cyber* OR Big data OR Artificial intelligence OR AI OR Industry 4.0 OR Smart*" AND "Review OR meta") As a result, 67 articles were included in group 1.
- **Group 2 Oft-cited articles.** Because no abstract analysis was done yet, the dataset remained 632 articles from the initial search. Following the standard set out by Crossan and Apaydin (2010), articles with at least five citations per year were chosen in group 2. This filter classified 419 articles in group 2.
- **Group 3 Residual articles/recent articles.** There were 213 residual articles that left from the filter in group 2. After a rough analysis, one possible explanation of these low citation rates was the recent publishing years, since they were mostly from 2016 to 2018.

The grouping result of the initial dataset is shown in Table 1.2.

Table 1.3 Summary of total identified articles for each filtering step

Groups/Number of articles	Group 1	Group 2	Group 3
Initial dataset	67	419	213
First filtering process	1	277	131
Second filtering process	1	136	63
Third filtering process	**0**	**30**	**9**

Data extraction

In order to select articles in line with the research objective, several filtering activities were done (results are shown in Table 1.3). To start, an abstract analysis was carried out for all the articles in group 1. As a result, no paper qualified. The excluded articles were either in the improper systematic literature standard or focused on customer review. We then proceeded to the data extraction for group 2 and group 3 by using three filtering processes: (1) screening the title, (2) screening the abstract, and (3) abstract and content analysis.

Irrelevant articles were removed after all the titles were screened. This filtering process mainly focused on the following aspects: (1) Out of focus topic. Articles focusing on societal behaviors (e.g., presidential election, online dating), human psychology, and consumer behaviors were excluded. (2) Irrelevant keywords (e.g., "Smartphone" from smart*). This filtering process excluded 142 articles from group 2 and 82 articles from group 3.

The second filtering process was about screening the abstracts. Since we focused on the impact of digitalization on business, we identified the keywords: design, structure, strategy, and organization as related to this topic. Therefore, abstracts that did not contain the above-mentioned keywords were excluded after screening. Additionally, articles that did not specifically focus on business were excluded. As a result, group 2 included 136 articles and group 3, 131.

The third filtering process combined an analysis of abstracts and contents. First, the abstracts of all the filtered articles were studied in order to make sure all the articles in the dataset were qualified within the research objective. We then focused on the contents. The selected articles focused on how companies/business deal with digitalization in terms of strategies and/or the company's structures. Articles about how digitalization affects business models were also included. Both technology providers and adaptors were taken into account. In the end, there were 30 articles left in group 2 and 9 articles left in group 3. In order to make the process of analysis easier, group 2 and group 3 were merged.

To extract the important data from the selected articles, a list of literature review questions (Table 1.4) was developed to classify the selected articles. The questions included the research aim, research target, research method, type of digitalization, technology mentioned, business range, and research contribution. The answers provided a preliminary understanding of the impact of digitalization on business and were used to further classifications.

Table 1.4 List of literature review questions

Conceptual decisions-related questions
What was/were the aim(s) of the research?
What was/were the research target(s)?
What was the research method?
Which types of digitalization are mentioned in the paper? (From 0 to 1 or Start from 1)
What was/were the technology(ies) mentioned in the research?
Which industry/types of businesses are research targets?
Which was/were the main contribution(s) of this research?

Results

Although all the selected articles focused on digitalization, they were embedded in disconnected literature streams. As a result, the 39 articles were categorized into 3 streams: (i) Big data and Big data analytics; (ii) Digital business strategy; and (iii) Digitalization in business. The overview is shown in Table 1.5.

Big Data (BD) and Big Data Analytics (BDA)

Big Data (BD) and Big Data Analytics (BDA) have greatly influenced traditional businesses. The concept of Data-Driven Business Models describes businesses that build upon data (Hartmann et al., 2016). Normally, Big Data is defined by the three Vs – Volume, Velocity, and Variety (Chen et al., 2015; Matthias et al., 2017) – but it has been broadened into more complex dimensions, including variability, veracity, and value (Matthias et al., 2017). Generally, Big Data is related to three main fields, namely, administrative, social media, and private sector data (Matthias et al., 2017).

Big Data Analytics is defined as "the application of statistical, processing, and analytics techniques to big data for advancing business" (Grover et al., 2018). Today, Big Data Analytics has developed and become essential for lots of businesses. With the increasingly advanced digital technologies, Big Data Analytics is able to capture and process a vast amount of data (Chen et al., 2015). Together, they create opportunities (summarized in Table 1.6) for business at both the corporate and supply chain levels (Chen et al., 2015; Johnson et al., 2017; Kache and Seuring, 2017).

Companies using Data-Driven Business Models use data as key sources, specifically: (i) raw data, (ii) processed data as information/knowledge, and (iii) non-data products or service (Hartmann et al., 2016).

Big Data and analytics can support businesses through different aspects. They can improve the decision-making process (Chen et al., 2015; Grover et al., 2018; Johnson et al., 2017), new product development process (Johnson et al., 2017), and information management (Kache and Seuring, 2017). They may also stimulate service innovation (Grover et al., 2018; Kache and Seuring, 2017; Lehrer

16 *Maria Carmela Annosi et al.*

Table 1.5 Category and number of papers in three streams

Streams	Number of papers	Reference
Big Data and Big Data Analytics	9	(Grover et al., 2018) (Lehrer et al., 2018) (Johnson, Friend, and Lee, 2017) (Troilo, De Luca, and Guenzi, 2017) (Gerard et al., 2016) (Chen et al., 2015) (Kache and Seuring, 2017) (Matthias et al., 2017) (Vidgen, Shaw, and Grant, 2017)
Digital business strategy	7	(Bharadwaj et al., 2013) (Adomavicius, Bockstedt, and Kauffman, 2008) (Woodard, Ramasubbu, and Sambamurthy, 2013) (Drnevich and Croson, 2013) (Granados and Gupta, 2013) (Grover and Kohli, 2013) (Mithas and Mitchell, 2013)
Digitalization in business	23	(Mehra et al., 2018) (Ramaswamy and Ozcan, 2018) (Toytari et al., 2018) (Fayard, Gkeredakis, and Levina, 2016) (Foucart, Wan, and Wang, 2018) (Benner, 2009) (Chen et al., 2012) (Austin, Devin, and Sullivan, 2012) (Ba, Stallaert, and Zhang, 2010) (Eaton et al., 2015) (Granados et al., 2008) (Khouja and Wang, 2010) (Bailey, Leonardi and Barley, 2012) (Blohm et al., 2016) (Dong, Xu and Zhu, 2009) (Feng, Guo and Chiang, 2009) (Granados and Gupta, 2013) (Henfridsson and Bygstad, 2013) (Lu et al., 2015) (Lusch and Nambisan, 2015) (Mariani, Di Felice, and Mura, 2016) (Marion, Barczak, and Hultink, 2014) (Marion, Meyer, and Barczak, 2015)

et al., 2018; Troilo et al., 2017) by enhancing customer experience through dynamic customization, increasing service process efficiency, and building service concepts (Troilo et al., 2017). Additionally, they create symbolic value, obtained through delivering an innovative image and/or maintaining one's reputation (Grover et al., 2018).

Similarly, Big Data and analytics can have positive impacts on operation and logistics in the supply chain (Kache and Seuring, 2017): they may increase

Table 1.6 The summary of opportunities brought by BD and BDA

Level	Opportunities	Reference
Corporate level	Decision making	(Johnson, Friend, and Lee, 2017) (Grover et al., 2018) (Chen, Preston, and Swink, 2015)
	New product development	(Johnson, Friend, and Lee, 2017)
	Service innovation	(Troilo, De Luca, and Guenzi, 2017) (Kache and Seuring, 2017) (Grover et al., 2018) (Lehrer et al., 2018)
	Operation	(Troilo, De Luca, and Guenzi, 2017) (Kache and Seuring, 2017) (Grover et al., 2018)
	Information management	(Kache and Seuring, 2017)
	Function and symbolic value	(Grover et al., 2018)
Supply chain level	Supply chain integration	(Johnson, Friend, and Lee, 2017) (Kache and Seuring, 2017)
	Supply chain performance	(Chen, Preston, and Swink, 2015)
	Visibility and transparency	(Kache and Seuring, 2017)
	Operation	(Kache and Seuring, 2017)
	Logistics	(Kache and Seuring, 2017)

visibility and transparency by providing real-time and end-to-end information sharing (Kache and Seuring, 2017) as well as increase asset productivity and business growth (Chen et al., 2015). Lastly, they support the integration of the supply chain (Johnson et al., 2017; Kache and Seuring, 2017). Supply chain collaboration and integration can build trust, which encourages information exchange (Kache and Seuring, 2017).

Nonetheless, some challenges exist. Access and reliability of these technologies are one of the problems (Matthias et al., 2017). Data quality is also essential for applying Big Data Analytics (Vidgen et al., 2017). Another challenge relates to people, specifically how they think, react, and respond to these new things (Matthias et al., 2017). Additionally, the business strategy and mission should fit with the implantation of technology since they decide the company's direction (Kache and Seuring, 2017).

Digital business strategy (DBS)

In the digital era, digital business strategy (DBS) has become popular in both the business and academic world. The definition of digital business strategy differs with scholars. In the research by Woodard et al. (2013), a digital business strategy is considered "a pattern of deliberate competitive actions undertaken by a firm as it competes by offering digitally enabled products or service." Bharadwaj et al. (2013) define digital business strategy as "organizational strategy formulated and executed by leveraging digital resources to create differential value." Furthermore, Mithas et al. (2013) indicate that digital business strategy is not only limited to the company and its surroundings, since it also has an impact on general digital business in the competitive environment. Nevertheless, they all agree on the importance of digital business strategy in both academic research and managerial practices.

With the development of information technology (IT), digitalization has become an irreversible trend and plays an important role in company strategy.

The differences between digital business strategy and traditional IT business strategy can be seen in the following four aspects: (1) scope, since digital business strategy is cross-functional within the larger business range; (2) scale, provided it also includes the digital aspects; (3) speed, as digitalization increases the pace of the business in several aspects, including product launches, decision making, supply chain orchestration, and network formation and adaptation; and, finally, (4) sources, such as information and multi-sided business models (Bharadwaj et al., 2013). These differences have generated interests among researchers about digital business strategy. In the research by Woodard et al. (2013), design moves (the separate strategic actions that can change the product or system of a company and can bring competitive advantages to the company) and design capital (the aggregate result of all the design moves) are identified as the basic logic of digital business strategy. Understanding the two concepts and the connections between them are beneficial for decision making in digital business strategy (Woodard et al., 2013).

Digitalization in Business

Digitization has decreased the overall profits of the traditional product supply chain, from producers to retailers (Khouja and Wang, 2010), since less physical activities and resources are required. We found that most contributions are related to the impact on operations, structure, innovation, and marketing.

The internal processes within the company have been affected greatly by the development of digital technologies. Social media and new media tools that create collaboration SPOT tools have been applied in new product development as enablers and facilitators of communication and knowledge exchange both inside and outside the organization (Marion, Barczak and Hultink, 2014). On the other hand, social networking tools have no direct relation with team collaboration, and worse, they appear to influence management evaluation negatively (Marion et al., 2014). IT infrastructure, the enabler of collaboration within the organization, encourages modular product architecture, while digital design tools like CAD do not work the same (Marion et al., 2015). These results indicate that companies have to be careful when implementing new media tools, since not all of them work the same.

Digitalization also appears in supply chains. Similarly, supply chain management (SCM) has started focusing on information-based integration along the chain (Zhu, 2004). However, the benefits rely on IT-related resources (Dong et al., 2009). IT contributes to supply chains mainly by enabling digital integration at the process level (Dong et al., 2009). Effectively using technology to improve upstream, downstream, and internal operations could improve the chain process, creating value (Dong et al., 2009).

Digitalization has also impacted company structure in terms of changing the way people work. A new type of work style – "smart" work – is created, which is free from geographic limitations. In the research by Bailey et al. (2012), the authors investigated three types of common virtual work, namely: virtual teams, remote control, and simulations. (1) Virtual teams: the improvement of digital communication tools enables people to interact and communicate from a distance, and representatives are not essential. (2) Remote control: different from virtual teams, remote control focuses more on objects instead of people. People manage the data collected through a physical system and then change the system through feedback. In this case, the object or the physical system serves as representatives. For these two types of virtual work, geographic limitations are broken by digital technology. On the other hand, we have (3) simulation, which focuses on study and experimentation instead of distance, although it can be overcome. As such, the company structure is greatly changed.

Knowledge management is also an area on which the impact of digitalization has been very strong. Knowledge management is essential for companies, it includes internal knowledge transferring and external knowledge generating. Culture might act as a moderator: for example, companies with different cultures would react differently to crowdsourcing, a tool to generate knowledge (Fayard et al., 2016), or to the use of corporate blogs (Lu et al., 2015). Similarly, process management practices also have an impact on companies' responses to new technology and vice versa (Benner, 2009). In fact, some management practices might cause distance between companies' capabilities and the changing environment, and as a result, some companies, especially incumbents, cannot react to rapid technological change (Benner, 2009).

The internet and mobile technologies have changed information transferring in the traditional market, and this change in information exposure has a twofold impact on markets (Granados and Gupta, 2013). On one hand, information sharing can reduce information asymmetry, while on the other, it can expose the company to its competitors (Grover and Kohli, 2013). Commonly, companies react through defensive denial (resistance to the inevitable trend) or have a passive reaction (simply waiting); however, these two actions are not long-term oriented; instead, companies should react with a transparency strategy (Granados and Gupta, 2013). The main concept behind the transparency strategy is to reveal information based on different parties through different strategic actions (disclose, distort, bias, or conceal) (Granados and Gupta, 2013).

Innovation is vitally important for company development since it can create a competitive advantage (Blohm et al., 2016; Fayard et al., 2016). The research by Austin et al. (2012) shows that digital technologies and digital systems are able to improve accidental innovation through different channels. First, random variations can be generated into the process, as well as opportunities to explore. Furthermore, though digital systems facilitate collections, random retrievals might also be created which encourage accidental innovation. Additionally, design systems might reduce the time to assess the results of each innovation iteration, so as to benefit the innovation process. Nevertheless, innovation does not always push technology forward; instead, it could create technological comebacks (Foucart et al., 2018).

Lastly, digitalization has created the brand-new e-market. Generally, IT can shake up traditional markets with three conditions: they are newly easy to enter, attractive to attack, and difficult to defend (Granados et al., 2008). These conditions happen mainly due to the low costs they bring thanks to IT. Starting in the 1990s, the internet has provided a new channel that has decreased the barrier to enter markets, with amazon.com in book markets being one example of this (Granados et al., 2008). When the customer, product, or activity profitability gradient comes along, markets are attractive to attack (Clemons, Gu, and Lang, 2002), and new entrants can target these markets with lower costs thanks to IT (Granados et al., 2008). In this context, even manufacturers are able to increase sales through adding an extra online channel because more potential clients can be reached, and search costs can be lower (Wu et al., 2008).

Digital channels have many advantages compared to physical retail shops, such as no rent for a physical space or to be opened 24/7. Traditional brick-and-mortar stores are also threatened by "showrooming," that is, when customers go to physical stores to observe or try the products but buy online at a lower price (Mehra et al., 2018). This free-riding behavior often leads to price competition and results in lower profits for both players (Mehra et al., 2018).

Digitization is also impacting services. Compared with traditional human-based services, digital services can decrease operating costs and improve service efficiency (Ba et al., 2010). With the development of digital technology, the concept of service has expended, in a logic of servitization. The digital service system is considered to be a complicated aggregation that consists of different players and resources (Eaton et al., 2015). Furthermore, smart services have been introduced that fundamentally change the values of a business (Toytari et al., 2018). Digital technologies play an important role in service innovation (Lusch and Nambisan, 2015), despite challenges such as those related to internal mindset and internal capability (Toytari et al., 2018). To deal with these barriers, the company must alter its mindset in terms of beliefs, norms, rules, and values, as well as capabilities including skills, routines, and assets (Toytari et al., 2018).

Conclusion

Digitalization has become an important topic in both the academic and business world. Our review presents some limitations. First, we have only extracted data from the Web of Science and restricted our analysis to articles published after 2008 and a limited range of keywords; nonetheless, previous literature has proposed relevant contributions to the topic and additional keywords may result in a wider set of articles to be analyzed. Since the research scope is quite large, regarding both digitalization and business, the number of selected articles was not large enough to provide an in-depth understanding of the impact. In fact, the selected 39 articles focus on different aspects in term of business.

Our review, though, points toward the idea that digitalization is an inevitable trend. Digital technology could bring competitive advantages by improving company performance in various aspects. However, it does not mean that the entire company can succeed or survive on digitalization, as underlined by the definition of the "digital vortex" (Bradley et al., 2015), representing the evolution of industries toward a "digital center" in which the business and activity will be totally digitalized, a phenomenon that will see four out of ten incumbents in major industries be forced out.

References

ABS, Chartered Association of Business Schools (2018). *The Academic Journal Guide*. London: Chartered Association of Business School – ABS.

Adomavicius, G., Bockstedt, J. C., Gupta, A., & Kauffman, R. J. (2008). Making sense of technology trends in the information technology landscape. *Mis Quarterly*, 32(4), 779.

Alaimo, C., & Kallinikos, J. (2017). Computing the everyday: social media as data platforms. *The Information Society*, 33(4), 175–191.

Austin, R. D., Devin, L., & Sullivan, E. E. (2012). Accidental innovation: supporting valuable unpredictability in the creative process. *Organization Science*, 23(5), 1505–1522.

Ba, S. L., Stallaert, J., & Zhang, Z. J. (2010). Balancing IT with the human touch: optimal investment in IT-based customer service. *Information Systems Research*, 21(3), 423–442.

Bailey, D. E., Leonardi, P. M. & Barley, S. R. (2012). The lure of the virtual. *Organization Science*, 23(5), 1485–1504.

Barrett, M., & Davidson, E. (2008). *Exploring the Diversity of Service Worlds in the Service Economy. In Information Technology in the Service Economy: Challenges and Possibilities for the 21st Century* (pp. 1–10). Boston, MA: Springer.

Benner, M. J. (2009). Dynamic or static capabilities? Process management practices and response to technological change. *Journal of Product Innovation Management*, 26(5), 473–486.

Bharadwaj, A., El Sawy, O. A., Pavlou, P. A., & Venkatraman, N. (2013). Digital business strategy: toward a next generation of insights. *MIS Quarterly*, 37(2), 471–482.

Blohm, I., Riedl, C., Fuller, J. & Leimeister, J. M. (2016). Rate or trade? Identifying winning ideas in open idea sourcing. *Information Systems Research*, 27(1), 27–48.

Bradley, J., Loucks, J., Macaulay, J., Noronha, A., & Wade, M. (2015). *Digital Vortex: How Digital Disruption is Redefining Industries*. Global Center for Digital Business Transformation: An IMD and Cisco initiative.

Calia, R. C., Guerrini, F. M., & Moura, G. L. (2007). Innovation networks: from technological development to business model reconfiguration. *Technovation*, 27(8), 426–432.

Chen, D. Q., Preston, D. S., & Swink, M. (2015). How the use of big data analytics affects value creation in supply chain management. *Journal of Management Information Systems*, 32(4), 4–39.

Christensen, C. M. (2007). *The Innovator's Dilemma: When New Technologies Cause Great Firms to Fail.* Brighton, MA: Harvard Business Review Press.

Clemons, E. K., Gu, B., & Lang, K. R. (2002). Newly vulnerable markets in an age of pure information products: an analysis of online music and online news. *Journal of Management Information Systems,* 19(3), 17–41.

Crossan, M. M., & Apaydin, M. (2010). A multi-dimensional framework of organizational innovation: a systematic review of the literature. *Journal of Management Studies,* 47(6), 1154–1191.

Dong, S. T., Xu, S. X., & Zhu, K. X. G. (2009). Information technology in supply chains: the value of IT-enabled resources under competition. *Information Systems Research,* 20(1), 18–32.

Drnevich, P. L., & Croson, D. C. (2013). Information technology and business-level strategy: toward an integrated theoretical perspective, *MIS Quarterly,* 37(2), 483–509.

Eaton, B., Elaluf-Calderwood, S., Sorensen, C., & Yoo, Y. (2015). Distributed tuning of boundary resources: the case of Apple's iOS service system. *MIS Quarterly,* 39(1), 217–243.

Fayard, A. L., Gkeredakis, E., & Levina, N. (2016). Framing innovation opportunities while staying committed to an organizational epistemic stance. *Information Systems Research,* 27(2), 302–323.

Feng, Y. F., Guo, Z. L., & Chiang, W. Y. K. (2009). Optimal digital content distribution strategy in the presence of the consumer-to-consumer channel. *Journal of Management Information Systems,* 25(4), 241–270.

Foucart, R., Wan, C., & Wang, S. (2018). Innovations and technological comebacks. *International Journal of Research in Marketing,* 35(1), 1–14.

Gerard, L., Camillia, M., & Linn, M. C. (2016). Technology as inquiry teaching partner. *Journal of Science Teacher Education,* 1(27), 1–9.

Gobble, M. M. (2018). Digitalization, digitization, and innovation. *Research-Technology Management,* 61(4), 56–59.

Granados, N., Gupta, A., & Kauffman, R. J. (2008). Designing online selling mechanisms: Transparency levels and prices. *Decision Support Systems,* 45(4), 729–745.

Granados, N., & Gupta, A. (2013). Transparency strategy: competing with information in a digital world. *MIS Quarterly,* 637–641.

Grover, V., & Kohli, R. (2013). Revealing your hand: caveats in implementing digital business strategy. *MIS Quarterly,* 37(2), 655–662.

Grover, V., Chiang, R. H., Liang, T. P., & Zhang, D. (2018). Creating strategic business value from big data analytics: a research framework. *Journal of Management Information Systems,* 35(2), 388–423.

Hartmann, P. M., Zaki, M., Feldmann, N., & Neely, A. (2016). Capturing value from big data – a taxonomy of data-driven business models used by start-up firms. *International Journal of Operations & Production Management,* 36(10), 1382–1406.

Henfridsson, O., & Bygstad, B. (2013). The generative mechanisms of digital infrastructure evolution. *MIS Quarterly,* 907–931.

Johnson, J. S., Friend, S. B., & Lee, H. S. (2017). Big data facilitation, utilization, and monetization: exploring the 3Vs in a new product development process. *Journal of Product Innovation Management,* 34(5), 640–658.

Kache, F., & Seuring, S. (2017). Challenges and opportunities of digital information at the intersection of Big Data Analytics and supply chain management. *International Journal of Operations & Production Management,* 37(1), 10–36.

Khouja, M., & Wang, Y. L. (2010). The impact of digital channel distribution on the experience goods industry. *European Journal of Operational Research*, 207(1), 481–491.
Lehrer, C., Wieneke, A., vom Brocke, J., Jung, R., & Seidel, S. (2018). How big data analytics enables service innovation: materiality, affordance, and the individualization of service. *Journal of Management Information Systems*, 35(2), 424–460.
Lu, B. J., Guo, X. H., Luo, N. L., & Chen, G. Q. (2015). Corporate blogging and job performance: effects of work-related and nonwork-related participation. *Journal of Management Information Systems*, 32(4), 285–314.
Lusch, R. F., & Nambisan, S. (2015). Service innovation: a service-dominant logic perspective, *MIS Quarterly*, 39(1), 155–175.
Mariani, M. M., Di Felice, M., & Mura, M. (2016). Facebook as a destination marketing tool: evidence from Italian regional Destination Management Organizations. *Tourism Management*, 54, 321–343.
Marion, T. J., Barczak, G., & Hultink, E. J. (2014). Do social media tools impact the development phase? An exploratory study. *Journal of Product Innovation Management*, 31(S1), 18–29.
Marion, T. J., Meyer, M. H., & Barczak, G. (2015). The influence of digital design and IT on modular product architecture. *Journal of Product Innovation Management*, 32(1), 98–110.
Matthias, O., Fouweather, I., Gregory, I., & Vernon, A. (2017). Making sense of Big Data – can it transform operations management? *International Journal of Operations & Production Management*, 37(1), 37–55.
Mehra, A., Kumar, S., & Raju, J. S. (2018). Competitive strategies for brick-and-mortar stores to counter "showrooming." *Management Science*, 64(7), 3076–3090.
Mithas, S., Tafti, A., & Mitchell, W. (2013). How a firm's competitive environment and digital strategic posture influence digital business strategy. *MIS Quarterly*, 37(2), 511–536.
Osterwalder, A., & Pigneur, Y. (2010). *Business model generation: a handbook for visionaries, game changers, and challengers*. Hoboken, NJ: John Wiley & Sons.
Porter, M. E., & Heppelmann, J. E. (2014). How smart, connected products are transforming competition. *Harvard Business Review*, 92(11), 64–88.
Ramaswamy, V., & Ozcan, K. (2018). Offerings as digitalized interactive platforms: a conceptual framework and implications. *Journal of Marketing*, 82(4), 19–31.
Scott, W. R. (1995). *Institutions and Organizations. Foundations for Organizational Science*. London: A Sage Publication Series.
Teece, D. J. (1980). Economies of scope and the scope of the enterprise. *Journal of Economic Behavior & Organization*, 1(3), 223–247.
Toytari, P., Turunen, T., Klein, M., Eloranta, V., Biehl, S., & Rajala, R. (2018). Aligning the mindset and capabilities within a business network for successful adoption of smart services. *Journal of Product Innovation Management*, 35(5), 763–779.
Tranfield, D., Denyer, D., & Smart, P. (2003). Towards a methodology for developing evidence-informed management knowledge by means of systematic review. *British Journal of Management*, 14(3), 207–222.
Troilo, G., De Luca, L. M., & Guenzi, P. (2017). Linking data-rich environments with service innovation in incumbent firms: a conceptual framework and research propositions *Journal of Product Innovation Management*, 34(5), 617–639.
Vidgen, R., Shaw, S., & Grant, D. B. (2017). Management challenges in creating value from business analytics. *European Journal of Operational Research*, 261(2), 626–639.

Woodard, C. J., Ramasubbu, N., Tschang, F. T., & Sambamurthy, V. (2013). Design capital and design moves: the logic of digital business strategy. *MIS Quarterly*, 37(2), 537–564.

Wu, D., Ray, G., & Whinston, A. B. (2008). Manufacturers' distribution strategy in the presence of the electronic channel. *Journal of Management Information Systems*, 25(1), 167–198.

Yoo, Y. (2012). Digital materiality and the emergence of an evolutionary science of the artificial. In P.M. Leonardi, A. Bonnie, and J. K. Nardi (Eds.), *Materiality and Organizing: Social Interaction in a Technological World*, 134–154.

Zhu, K. (2004). Information transparency of business-to-business electronic markets: a game-theoretic analysis. *Management Science*, 50(5), 670–685.

Zott, C., & Amit, R. (2010). Business model design: an activity system perspective. *Long Range Planning*, 43(2–3), 216–226.

2 Digital technology in the agri-food sector

A review on the business impact of digitalization in the agri-food sector

Maria Carmela Annosi and Federica Brunetta

Introduction

Organizations in all fields are now increasingly making use of digital technologies, which has resulted in a new phenomenon defined as "Industry 4.0."

Following the "4.0 era," the agricultural industry is also on the verge of a digital transformation, with a growing number of organizations, both established and start-ups, and investors devoting enormous resources to R&D, adoption, and diffusion of new technologies in agriculture (Deloitte, 2016), resulting in "Agrifood 4.0" (Miranda et al., 2019). "Digital agriculture" is defined by Shepherd et al. (2018) as "the use of detailed digital information to guide decisions along the agricultural value chain." Within agriculture, the use of Precision Agriculture Technologies (PAT), or Smart Farming Technologies, are mostly applied on the farm in the form of digital input. Nonetheless, Agrifood 4.0 innovations can apply to different actors in the value chain, as in the case of marketplace technologies. Smart technologies and 4.0 are showing extensive benefits, with a potential large impact both on economics and on social and environmental issues since it is enhancing the efficiency of operations by reducing the overuse of inputs (i.e., water, fertilizer, pesticides, seeds, etc.). When applied to several actors in the value chain, these technologies consistently reduce waste and costs, be it on the farm, in retail, or in consumption, as they provide new and valuable information for decision making (FAO, 2013). Within the food industry, the adoption of robots in the production line and the use of automations have become very attractive in light of the drop-in production costs. Adopting these technologies will be necessary for an industry like food, which has vast competition and significantly lower labor costs abroad (Masey et al., 2010).

While effects on the whole value chain are important, we focus specifically on agriculture. Despite the significant benefits and the efforts of many institutions to pave the way for Smart Agriculture by providing consistent public funds and introducing policies to support innovation, there are still many farms coping with challenges in adopting digitalization. This chapter focuses on the digitalization of agriculture and refers to existing literature to summarize the main challenges faced by agri-food firms when adopting digitalization. By not making use of new technologies while other firms do, a business might fall behind amongst the vast competition (Rao, 2003). Additionally, fast growing industries are moving on to a "5.0 era," while the agricultural sector still has difficulties in adopting "Industry 4.0" technologies (Zambon et al., 2019).

Based on these earlier studies, there are concerns about the speed of adopting digital technologies in the agricultural sector and the lack of a clear overview of the problems that agricultural firms face in adopting digitalization in their operations. Moreover, there is also a relevant research gap that needs to be examined and which will be essential for decision makers in this area, and given its immense potential economic and environmental impact, studying the factors affecting technology adoption is crucial.

Digital technologies in agri-food

In Chapter 1, we highlighted the main trends related to digitalization, which are extending to the agricultural industry driven primarily by four megatrends: First, the need for sustainable production as well as to find alternative and innovative ways to feed the growing global population, which is expected to increase to 8 billion by 2025 and 9.7 billion by 2050 (FAO, 2013). Second, consumer-specific demand for safety, traceability, ecological footprint, and health properties of food, as demonstrated by the rapid growth of the organic and slow food market (IFOAM EU Group, 2016). Third, the emphasis on potentially polluting effects, biodiversity loss, and reduction of soil fertility in relation to agri-food operations, as well as the quest to reduce the impact of climate change and to increase the optimization of resources (e.g., water and land scarcity following urbanization) (World Business Council for Sustainable Development (WBCSD), 2008). Lastly, the technological shift that has enabled the development of agri-food applications. These include the explosion of smartphone capabilities and the diffusion of the internet and Wi-Fi, which serve as catalysts for this change, as well as the development of technologies related to Artificial Intelligence (AI), Big Data Analytics, Cloud Computing, Cyber-Physical Systems (CPS) (mechanisms controlled or monitored by computer-based algorithms), and the Internet of Things (IoT). More specifically, in this chapter, when discussing "Digital Technologies in Agri-food" or "Agriculture 4.0," we are referring to the application of such technologies in farming and the implementation of Smart Farming (also defined as Precision Farming), following the "Fourth Agricultural Revolution" (Deloitte, 2016). We describe the main technologies below:

- *Smart Greenhouses*: self-regulating, climate-controlled environments for optimal horticulture growth with minimal human intervention. Within a smart greenhouse, moisture, humidity, and light are constantly monitored and any divergence from the predetermined conditions is automatically adjusted.

- *Smart Irrigation Control Systems*: IoT and AI systems are used to minimize water usage in irrigation. Such systems combine technologies monitoring soil moisture, wind, and rain, as well as remote sensing controllers and AI algorithms (Shitu et al., 2015). Thus, Smart Irrigation optimizes water consumption according to soil data and through variable rate technologies (VRT).
- *Drones and Robots*: the former is an unmanned aerial vehicle used in farming to help monitor crop growth. Farmers can monitor fields from the sky and gather richer pictures through sensors and digital imaging, which is helpful for detecting potential problems in cultivation, pests, and other crop diseases. Robots, on the other hand, are "intelligent farm machines" that automatize and perform diverse farm tasks and can be remotely controlled.
- *Soil, Plants, and Yield Monitoring Systems*: in these systems, several smart sensors are used to gather data, increase efficacy, and prevent problems. They are used to monitor various physical, chemical, and biological soil properties, plant compositions, mass flow, and harvest.
- *Software and Data Analytics*: following the need for data-driven insights, organizations might use software to track, manage, and maximize the use of resources and production, as well as data analytics, information collection, and management to gather not only real-time insights but, more importantly, predictions. Software can also be used to increase collaborations with other actors along the supply chain.
- *Precision Livestock*: just like in the case of precision farming, technologies can be used to optimize operations and deliver better results in livestock farming. For example, technologies allow farmers not only to monitor the health and welfare of livestock, but also feeding, heat stress, milk harvest, and breeding patterns.
- Lastly, technologies can act as enablers and drivers for commercialization by providing *marketplaces*, connecting organizations to suppliers or consumers, even bypassing intermediation.

Methodology

In order to collect data about the challenges related to digitalization in the agri-food industry, we performed a systematic review of the literature, following the method laid out by Tranfield et al. (2003) described in Chapter 1 (Par "Systematic literature review: the methodology" in Chapter 1, ibidem).

We performed the research with Scopus, using search terms associated with the 4.0 era (digit*, big data, Artificial intelligence) in relation to "agri-food," "agriculture," "farm*," and "smart farm*." We restricted the search to the "social sciences," journals, peer reviewed articles, and documents written in English. Additional filters included the number of citations (if the paper did not fit in any of the three groups described below) and outlet (only ABS 2, 3, 4 and 4* journals). We followed the grouping method described by Crossan and Apaydin (2010) and divided the papers into three groups: the first group containing 28 reviews and meta-analysis, the second group consisting of 47 frequently cited papers (at least five times per year), and the third group comprising 433 more recent papers (published between 2009 and 2019). We excluded review articles and filtered the remaining ones. We were left with 17 papers after filtering, most of which were empirical, summarized in Table 2.1.

Discussion

We analyzed the papers to identify the challenges faced by agricultural firms. We started by classifying the challenges highlighted by the different papers and the theories connected to each type of challenge. Based on the available literature, several features could be significant in the adoption and diffusion of a technological innovation, some of which are found at the individual level, like a farmer's educational background and his/her capability to perceive an opportunity. Others are at the firm or environmental level, and relate to factors such as finance, access to physical infrastructure and business services, institutional support, and the socio-cultural context (Long, Block and Poldner., 2017; Annosi et al., 2019).

Individual factors, such as education and cognitive capacity, may influence the decision to adopt and use a technology. These capabilities allow an individual to identify a match between the opportunities and threats and internal resources, skills, and capabilities (Shane and Venkataraman, 2000), since they relate to the ability to evaluate benefits and costs, identify needs and opportunities (i.e., turning a farm into a "smart" one) (DeTienne and Chandler, 2004), and act on them, using the technologies (Annosi et al., 2019). In this light, education and training enhance an individual's own cognitive abilities to identify the quantity and characteristics of technologies to adopt (Fernandez-Cornejo, Beach, and Huang, 1994; Daberkow and McBride, 2003; Paxton et al., 2010). At the same time, farmers who perceive net benefits and possess the education and training needed to use PAT showed a greater propensity to adopt them (Adrian et al., 2005). An additional role of cognitive capacity might be the one played by socio-cultural norms and culture, such as peer pressure, which might drive or hinder decisions to adopt Smart Agriculture. For example, in his study of innovation adoption in India, Abdullah (2015) identified elements such as access, level of education, knowledge, quality, and landholding as potential features that have an impact on adoption. Specifically, he analyzed the level of education and landholding as proxies for caste to verify how individual and socio-cultural factors influenced the adoption and use of farming technology. Several of the analyzed papers focus at these cognitive factors. Bello-Bravo et al. (2018) built a study to verify the impact of diverse learning tools on rural populations. While the scope of their experiments ranges from health to agricultural issues, their approach highlights some interesting insights into factors such as access, education, costs of Information and Communication Technology (ICT) transfer, and learning. Chandra et al. (2017) analyzed climate-resiliency field schools (where the practice of organic farming and community seed bank

Table 2.1 List of articles analyzed

Year	Authors	Title	Journal	Themes
2003	Cecchini, S. and Scott, C.	Can information and communications technology applications contribute to poverty reduction? Lessons from rural India	*Information Technology for Development*	Access; Infrastructure; Institution
2009	Richards, P., de Bruin-Hoekzema, M., Hughes, S. G., Kudadjie-Freeman, C., Kwame Offei, S., Struik, P. C., and Zannou, A.	Seed systems for African food security: linking molecular genetic analysis and cultivator knowledge in West Africa	*Technology*	Institutions; Policy
2010	Mokotjo, W., and Kalusopa, T.	Evaluation of the Agricultural Information Service (AIS) in Lesotho	*International Journal of Information Management*	Access; Services
2011	Islam, M. S., and Grönlund, Å.	Bangladesh calling: farmers' technology use practices as a driver for development	*Information Technology for Development*	Age and Generation
2011	Soomai, S. S., Wells, P. G., and MacDonald, B. H.	Multi-stakeholder perspectives on the use and influence of "grey" scientific information in fisheries management	*Marine Policy*	Complexity; Services
2014	Hay, R., and Pearce, P.	Technology adoption by rural women in Queensland, Australia: women driving technology from the homestead for the paddock	*Journal of Rural Studies*	Access; Age and Gender; Education
2015	Abdullah, A.	Digital divide and caste in rural Pakistan	*The Information Society*	Access; Level of Education; Infrastructure; Socio-factors
2015	Tanure, S., Nabinger, C., and Becker, J. L	Bioeconomic Model of Decision Support System for farm management: proposal of a Mathematical Model	*Systems Research and Behavioral Science*	Complexity

(continued)

Table 2.1 Cont.

Year	Authors	Title	Journal	Themes
2016	Hennessy, T., Läpple, D., and Moran, B.	The digital divide in farming: a problem of access or engagement?	*Applied Economic Perspectives and Policy*	Access; Farm characteristics
2017	Chandra, A., Dargusch, P., McNamara, K. E., Caspe, A. M., and Dalabajan, D.	A study of climate-smart farming practices and climate-resiliency field schools in Mindanao, the Philippines	*World Development*	Access; Knowledge
2017	Panagiotopoulos, P., Bowen, F., and Brooker, P.	The value of social media data: integrating crowd capabilities in evidence-based policy	*Government Information Quarterly*	Knowledge; Policy
2017	Pant, L. P., and Hambly Odame, H.	Broadband for a sustainable digital future of rural communities: a reflexive interactive assessment	*Journal of Rural Studies*	Access
2018	Bello-Bravo, J., Tamò, M., Dannon, E. A., and Pittendrigh, B. R.	An assessment of learning gains from educational animated videos versus traditional extension presentations among farmers in Benin	*Information Technology for Development*	Access; Education; Knowledge
2018	Coble, K. H. Mishra, A. K., Ferrell, S., and Griffin, T.	Big data in agriculture: a challenge for the future	*Applied Economic Perspectives and Policy*	Access; Data Management; Policy
2018	Khanna, M., Swinton, S. M., and Messer, K. D.	Sustaining our natural resources in the face of increasing societal demands on agriculture: directions for future research	*Applied Economic Perspectives and Policy*	Incentives; Finance; Institutions
2018	Saggi, M. K., and Jain, S.	A survey toward an integration of big data analytics to big insights for value-creation	*Information Processing and Management*	Complexity; Finance
2019	Rotz et al.	Automated pastures and the digital divide: how agricultural technologies are shaping labor and rural communities	*Journal of Rural Studies*	Institutions; Policy

establishments are taught) and noticed that since climate-smart interventions are traditionally knowledge-intensive processes, a lack of education and previous experience can hamper the adoption of such technologies. One peculiar finding is that of Islam and Grönlund (2011), who, in their research about farmers in Bangladesh, showed that education and income did not represent barriers to adoption, but rather demographic factors such as age or having young children were.

These innovations are often costly and complex (Tanure et al., 2015; Saggi and Jain, 2018). Therefore, not only do individual factors play a role, but so do factors at the firm or environmental level, which can have an impact on the likelihood of innovation adoption and usage (Annosi et al., 2019). Among them, access to finance, institutional support, infrastructure and services, and cultural and social context. With regards to finance and costs, it is important to note that the higher the effective – or perceived – cost of digital technologies in the agri-food sector and the more difficulties there are in accessing financing, the lower the probability that farmers, entrepreneurs, or firms will adopt and use such technologies. Richards et al. (2009) accurately underlined this problem when they analyzed the challenges faced by African countries in terms of lack of support in funding and direct links between farmers and researchers (Richards et al., 2009). Institutions also play a role, especially through policies that stimulate farmers' investments in digital solutions or help them use such technologies (Khanna et al., 2018; Rotz et al., 2019). Another relevant issue, given the nature of these technologies, is the access to infrastructure, such as broadband internet and the cloud, which are required for Smart Agriculture to be used effectively and offer successful results. This was highlighted by Pant et al. (2017) in their study of examining Canadian farms. Similarly, services in commercial, legal, financial, and, most importantly, IT and digital consultancy could offer proper support to farmers deciding to adopt certain innovations (Hay and Pearce, 2014; Long et al., 2016). Mokotjo and Kalusopa (2010) noticed that the use of agricultural information services is related to weak promotion and training in the use of these services. For example, Soomai et al. (2011), in their study of fisheries, noticed that institutional support could help entrepreneurs cope with the high technical content. From a different perspective, Panagiotopoulos et al. (2017) noticed that in the UK, farmers have also successfully used digital platforms to influence policy making. Other authors specifically focus on these issues. For instance, Cecchini and Scott (2003), focusing on how technologies support farmers in rural India and their connection to markets, analyzed the digital divide and identify access to a proper infrastructure as a necessary, but not sufficient, prerequisite for adopting and using technologies, recalling the importance of incentives and support. Coble et al. (2018) also mentioned how infrastructure is a critical bridge in the use of technology. Thus, access and availability of an infrastructure result in a comparative advantage for firms that have it. Hay and Pearce (2014) confirmed this in their study of the lifestyle of rural women in Queensland.

At the firm level, the adoption of smart farming technologies was found by Hennessy et al. (2016) to be dependent on a business's characteristics and not simply on the access to digital technologies; indeed, in their study, farmers with access to computers did not necessarily adopted ICT technologies.

Conclusion

In this chapter, we provided an overview of existing literature to assess the challenges previously highlighted regarding the adoption of digitalization in agriculture.

While there has been abundant literature on Precision Agriculture (e.g., Cox, 2002; Grogan, 2012; Kaloxylosab et al., 2012), it is evident that the contributions to management literature related to digitalization and agri-food are still limited, and a great deal remains to be done to further understand the factors that influence farmers' decision to adopt smart solutions. So far, management scholars have focused on the impact of new technologies on a farm's business model (e.g., Long, Blok, and Poldner, 2017) or firm performance following adoption, rather than focusing on the relationship between the decision to invest in smart technologies and the various factors that impact such a decision (Annosi et al., 2019). Nonetheless, despite the scope and variety of topics analyzed, the purpose of this chapter is to lay the groundwork for the following chapters, where some of these issues are discussed in detail.

References

Abdullah, A. (2015). Digital divide and caste in rural Pakistan. *Information Society*, 31(4), 346–356.

Adrian, A. M., Norwood, S. H., & Mask, P. L. (2005). Producers' perceptions and attitudes toward precision agriculture technologies. *Computers and Electronics in Agriculture*, 48, 256–271.

Annosi, M. C., Brunetta, F., Monti, A., & Nati, F. (2019). Is the trend your friend? An analysis of technology 4.0 investment decisions in agricultural SMEs. *Computers in Industry*, 109, 59–71.

Bello-Bravo, J., Tamò, M., Dannon, E. A., & Pittendrigh, B. R. (2018). An assessment of learning gains from educational animated videos versus traditional extension presentations among farmers in Benin. *Information Technology for Development*, 24(2), 224–244.

Cecchini, S., & Scott, C. (2003). Can information and communications technology applications contribute to poverty reduction? Lessons from rural India. *Information Technology for Development*, 10, 73–84.

Chandra, A., Dargusch, P., McNamara, K. E., Caspe, A. M., & Dalabajan, D. (2017). A study of climate-smart farming practices and climate-resiliency field schools in Mindanao, the Philippines. *World Development*, 98, 214–230.

Coble, K. H., Mishra, A. K., Ferrell, S., & Griffin, T. (2018). Big data in agriculture: a challenge for the future. *Applied Economic Perspectives and Policy*, 40(1), 79–96.

Cox, S. (2002). Information technology: the global key to precision agriculture and sustainability. *Computers and Electronics in Agriculture*, 36(2–3), 93–111.

Crossan, M. M., & Apaydin, M. (2010). A multi-dimensional framework of organizational innovation: a systematic review of the literature. *Journal of Management Studies*, 47(6), 1154–1191.

Daberkow, S. G., & McBride, W. D. (2003). Farm and operator characteristics affecting the awareness and adoption of precision agriculture technologies in the US. *Precision Agriculture*, 4(2), 163–177.

Deloitte. (2016). *From Agriculture to AgTech: an industry transformed beyond molecules and chemicals*. Retrieved from www.gita.org.in/Attachments/Reports/Deloitte-Tranformation-from-Agriculture-to-AgTech.pdf

DeTienne, D., & Chandler, G. (2004). Opportunity identification and its role in the entrepreneurial classroom: a pedagogical approach and empirical test. *Academy of Management Learning and Education*, 3(3), 242–257.

FAO. (2012). *World Agriculture towards 2030/2050: the 2012 Revision*. In: FAO (Ed.). Food and Agricultural Organization of the United Nations, Rome, Italy.

FAO. (2013) Climate-smart Agriculture – Sourcebook in Nations. In: FAO (Ed.). *Food and Agricultural Organization of the United Nations*, Rome, Italy.

Fernandez-Cornejo, J., Beach, E. D., & Huang, W. J. (1994). The adoption of IPM techniques by vegetable growers in Florida, Michigan and Texas. *Journal of Agricultural & Applied Economics*, 26(1), 158–172.

Grogan, A. (2012). Smart farming. *Engineering & Technology*, 7(6), 38.

Hay, R., & Pearce, P. (2014). Technology adoption by rural women in Queensland, Australia: women driving technology from the homestead for the paddock. *Journal of Rural Studies*, 36, 318–327.

Hennessy, T., Läpple, D., & Moran, B. (2016). The digital divide in farming: a problem of access or engagement? *Applied Economic Perspectives and Policy*, 38(3), 474–491.

IFOAM EU Group. (2016). *Organic in Europe, Prospects and Developments 2016*. Retrieved from www.ifoam-eu.org/sites/default/files/ifoameu_organic_in_europe_2016.pdf

Islam, M. S., & Grönlund, Å. (2011). Bangladesh calling: farmers' technology use practices as a driver for development. *Information Technology for Development*, 17(2), 95–111.

Kaloxylosab, A., Eigenmannc, R., Teyed, F., Politopouloue, Z., Wolfertf, S., … Kormentzase, G. (2012). Farm management systems and the future internet era. *Computers and Electronics in Agriculture*, 89, 130–144.

Khanna, M., Swinton, S. M., & Messer, K. D. (2018). Sustaining our natural resources in the face of increasing societal demands on agriculture: directions for future research. *Applied Economic Perspectives and Policy*, 40(1), 38–59.

Long, T. B., Blok, V., & Coninx, I. (2016). Barriers to the adoption and diffusion of technological innovations for climate-smart agriculture in Europe: evidence from the Netherlands, France, Switzerland and Italy. *Journal of Cleaner Production*, 112, 9–21.

Long, T. B., Blok, V., & Poldner, K. (2017). Business models for maximising the diffusion of technological innovations for climate-smart agriculture. *International Food and Agribusiness Management Review*, 20(1030-2017-2134), 5–23.

Masey, R. J. M., Gray, J. O., Dodd, T. J., & Caldwell, D. G. (2010). Guidelines for the design of low-cost robots for the food industry. *Industrial Robot*, 37(6), 509–517.

Miranda, J., Ponce, P., Molina, A., & Wright, P. (2019). Sensing, smart and sustainable technologies for Agri-Food 4.0. *Computers in Industry*, 108, 21–36.

Mokotjo, W., & Kalusopa, T. (2010). Evaluation of the Agricultural Information Service (AIS) in Lesotho. *International Journal of Information Management*, 30(4), 350–356.

Panagiotopoulos, P., Bowen, F., & Brooker, P. (2017). The value of social media data: integrating crowd capabilities in evidence-based policy. *Government Information Quarterly*, 34(4), 601–612.

Pant, L. P., & Hambly Odame, H. (2017). Broadband for a sustainable digital future of rural communities: a reflexive interactive assessment. *Journal of Rural Studies*, 54, 435–450.

Paxton, K. W., Mishra, A. K., Chintawar, S., Larson, J. A., Roberts, R. K., English, B. C., & Lambert, D. M. (2010). *Precision Agriculture Technology Adoption for Cotton Production*. Griffin, GA: Southern Agricultural Economics Association.

Rao, N. H. (2003). Electronic commerce and opportunities for agribusiness in India. *Outlook on Agriculture*, 32(1), 29–33.

Richards, P., de Bruin-Hoekzema, M., Hughes, S. G., Kudadjie-Freeman, C., Kwame Offei, S., Struik, P. C., & Zannou, A. (2009). Seed systems for African food security: linking molecular genetic analysis and cultivator knowledge in West Africa. *Technology*, 45, 196–214.

Rotz, S., Gravely, E., Mosby, I., Duncan, E., Finnis, E., Horgan, M., ... Fraser, E. (2019). Automated pastures and the digital divide: how agricultural technologies are shaping labour and rural communities. *Journal of Rural Studies*, 68, 112–122.

Saggi, M. K., & Jain, S. (2018). A survey towards an integration of big data analytics to big insights for value-creation. *Information Processing and Management*, 54(5), 758–790.

Shane, S., & Venkataraman, S. (2000). The promise of entrepreneurship as a field of research. *Academy of Management Review*, 25(1), 217–226.

Shepherd, M., Turner, J. A., Small, B., & Wheeler, D. (2018). Priorities for science to overcome hurdles thwarting the full promise of the "digital agriculture" revolution. *Journal of the Science of Food and Agriculture*, 1–10.

Shitu, G. A., Maraddi, G. N., & Sserunjogi, B. (2015). A comparative analysis in resource utilization and yield performance of precision farming technologies in North Eastern Karnataka. *Indian Journal of Economics and Development*, 11(1), 137–145.

Soomai, S. S., Wells, P. G., & MacDonald, B. H. (2011). Multi-stakeholder perspectives on the use and influence of "grey" scientific information in fisheries management. *Marine Policy*, 35(1), 50–62.

Tanure, S., Nabinger, C., & Becker, J. L. (2015). Bioeconomic Model of Decision Support System for farm management: proposal of a Mathematical Model. *Systems Research and Behavioral Science*, 32(6), 658–671.

Tranfield, D., Denyer, D., & Smart, P. (2003). Towards a methodology for developing evidence-informed management knowledge by means of systematic review. *British Journal of Management*, 14(3), 207–222.

World Business Council for Sustainable Development (WBCSD). (2008). *Agricultural Ecosystems Facts and Trends*. Retrieved from http://cmsdata.iucn.org/downloads/agriculturalecosystems_2.pdf

Zambon, I., Cecchini, M., Egidi, G., Saporito, M. G., & Colantoni, A. (2019). Revolution 4.0: industry vs. agriculture in a future development for SMEs. *Processes*, 7(1), 36.

Part II
Challenges and current strategies in digitalization in agri-food

3 The role of managers or owners of SMEs in driving the digitalization process in the agri-food sector

Ivan Bedetti, Maria Carmela Annosi, Giorgia Bucci, Deborah Bentivoglio, Wilfred Dolfsma, and Adele Finco

Introduction

The adoption of digital technologies, also referred to as digital transformation, is a challenging process of change that has to deal with strategy more than technology per se (Kane et al., 2015). Due to a unique position that allows them to shape strategy, extant research has demonstrated that top managers influence the choices related to adopting technology (Damanpour and Schneider, 2008; Jarvenpaa and Ives, 1991; Midavaine et al., 2016; Rizzoni, 1991; Shah Alam, 2009). Moreover, top management patronage and behavior are critical for creating a supportive climate as well as providing suitable resources (Low et al., 2011; Walsh, 1988). Extant literature investigating top managers' intention to adopt digital technologies has been focusing on variables such as education, age, gender, etc. (Bantel and Jackson, 1989; Young et al., 2001). Yet, these variables may only partially explain differences in adoption behaviors. Since choices within the organization reflect the top management's cognitive elements and values, the way top managers behave toward championing innovations may function as an intermediary amongst the environment and the assimilation of the technology within the organization (Lin et al., 2014). Therefore, understanding the attitudes, motivations, characteristics, values, and subsequent behavior of the manager is central to understanding how and why technological innovation is accepted and implemented by some firms and not by others.

Managerial influence on strategic technological change is even more emphasized at the small and medium-sized enterprise (SME) level, where the top manager or owner–manager is considered to be an "all-rounder," involved in every organizational process and playing a unique position in the firm's decision-making process (Berergon and Raymond, 1992; Geletkanycz and Hambrick, 1997; Hambrick and Mason, 1984; Jeyaraj et al., 2006; Winston and Dologite, 2002). At the SME level, the adoption of digital technology was studied in terms of factors influencing the firm, either externally, such as competitive pressure and network influence, or internally, such as technological competences and human capital (Giotopoulos et al., 2017; Mehrtens et al.,

2001; Nguyen, 2009). A similar approach was applied to SMEs in the agri-food sector, and among the factors considered to influence the firm in the adoption process, there was evidence of top management's support (Premkumar and Roberts, 1999).

More recently, new theoretical approaches were adopted to better analyze how managers promote and sustain a successful digital transformation at the SME level (Li et al., 2018; Warner and Wäger, 2019). According to Teece (2017), the dynamic capabilities can be divided into three categories: sensing, seizing, and transforming. They are used to indicate different activities performed by the manager, such as identifying technological opportunities, mobilizing resources to address specific needs and opportunities, and carrying out continuous renewal.

Based on the work by Lin et al. (2014) and to pursue the research stream presented by Annosi et al. (2019) regarding the role of the manager in the process of adopting digital technologies by agri-food SMEs, we have developed this qualitative study.

Methods

To investigate how the managers of agri-food SMEs actively promote and sustain the successful adoption and usage of digital technology, the Grounded Theory (GT) methodology was applied in this study. According to Charmaz and Belgrave (2012), GT is a systematic inductive methodology to conduct qualitative research aimed toward developing new theories. Since research and theory on the subject of technology diffusion in this sector are still emerging, this study presents a multiple case study approach, selected because of its particular strengths in developing and extending theory (Eisenhardt and Graebner, 2007; Yin, 2003). The context of the agri-food sector seems to be attractive for studying the role of the manger in digital technology adoption and usage for several reasons. First of all, SMEs face several obstacles linked to the introduction of technological innovations, such as a lack of human and financial resources, limited organizational capabilities, and lack of a strategic vision (Kuan and Chau, 2001; Mehrtens et al., 2001; Premkumar, 2003). Secondly, these weaknesses the SMEs typically have are even more emphasized in the agri-food sector, where they struggle not only because of limited resources but also because of the lack of an appropriate digital infrastructure and training (Baourakis et al., 2002; Pickernell et al., 2004).

Based on these premises, data were collected from a sample of 24 agri-food SMEs in the Marche region (Italy) through semi-structured interviews during the period of March–April 2019. All the interviewees covered a managerial position (CEO, owner). We focused on the manager or owner of the enterprises because in SMEs, these people assume a central position and are considered to be the ones who make the final decision related to the internal and external organization of the firm (Bridge et al., 2003; Durst and Runar Edvardsson, 2012). The firms were selected based on their experience in the adoption and usage of digital technologies. The process of sampling concluded when the saturation point was attained, a condition in which a newly added unit of analysis did not provide any additional relevant information. The saturation point was reached after 33 interviews for a total of 24 SMEs. Then, the total sample was divided into two subgroups, creating two different scenarios of adoption and usage. The first group of Low Technology Integration (LTI) enterprises makes use of digital technologies only to a limited extent and solely concerning administration. The second group, on the other hand, comprising High Technology Integration (HTI) enterprises, vaunts a greater inclusion of digital technologies, showing increased usage both in the areas of administration and production. Additionally, a cross-case comparison between the two groups was made, focusing on the differences that emerged to increase the robustness and the generalizability of the resulting theory.

Data were analyzed based on the canons and procedures of GT research (Corbin and Strauss, 1990). In particular, it was used as an iterative coding process between the data collected, the literature found, and the grounded categories of concepts that emerged from the analysis. The coding process was repeated three times, advancing from simple and evident patterns to crosscutting themes and insights from the theory (Gioia et al., 2013). Each round of coding was done independently by the two researchers to increase the reliability of the study. What emerged from the comparison of the two groups of firms is a set of specific beliefs, intuitions, behaviors, and practices that are only present in the group of companies that undergo a successful process of digital transformation, the HTI enterprises. These findings are explained in more detail in the following sections.

Findings

In the following sections, we describe the theoretical themes related to managerial characteristics that emerged from the corresponding grounded model by comparing examples of codes emerging from LTI and HTI firms. To support our arguments, we provide quotes taken from the interview transcripts as explicative examples.

Managerial characteristics

Managerial beliefs

The strong belief of the HTI firms' managers in technology as a mean to innovate and improve their business is the first feature that could explain the difference between the HTI and LTI enterprises. In fact, more so than all the other managerial characteristics, discussed later on in this section, this belief is considered one of the most relevant factors affecting the adoption of technology since it represents the starting point on which decisions, and consequently actions and practices, are based (Lefebvre et al., 1997; Tikkanen et al.,

2005). Intuitively, we observed that HTI firms' managers strongly believe in the adoption of new technologies as a means through which the firm adapts to growing competitive pressure and new market requirements (Bertrand and Schoar 2003; Davis, 1989). The dominant managerial belief that emerged from the interviews with HTI firms' managers is that improving enterprise performance strongly depends on technology adoption, as stated by one manager of Company 7:

> Performance improvement passes through technology adoption.

Company 7 is the highest technological integrated enterprise among our sample. Therefore, this belief is the consequence of a considerable application of technologies in the business. We understood from the interview that this belief is the result of a process of awareness that started in a moment of generational change, when the company was transferred from father to son. The need for innovation became evident for the young, new owners of Company 7 after they realized that only the integration of new technology in their business could solve the complexities and obligations that had drastically increased since their Father's Day.

On the contrary, even if some managers of LTI enterprises confirmed the importance of technology as a driving factor in their long-term strategy, we observed discrepancies between the belief and the actual behaviors and actions that were not directed toward technological innovation. For instance, the manager of Company 2 decided to adopt certain kind of digital technology only because it was mandatory:

> We adopted software for electronic invoicing [...], partly because it is now mandatory.

Even though he affirmed the relevance of technological innovations in his business, the manager of Company 1 perceived the use of technology as a threat:

> I see technology as an additional interference of a certain kind of industry in the agricultural sector.

Both these two LTI firms are characterized by a more traditional and conservative way of doing business. This type of business management derives from the basic belief that technology does not lead the firm's performance but is only considered as a marginal factor in the business strategy.

Furthermore, we observed that some LTI firms' managers complained about the current technological underdevelopment among the different players of the local agri-food supply chain. It is not uncommon that in the agri-food chain, some farmers are not using any digital technology at all; in some cases, they don't even have a computer to send and receive e-mails. Within the context of digital technology adoption, this awareness stands as a barrier. Here is one example of this situation explained by the manager of Company 22:

> let's just say that our farms, which is to say our suppliers, started to use e-mails only last year because they were obliged by the government, which had imposed electronic invoicing. Before last year, I was forced to send them all communications by post. Do you understand where the problem is in our business?

This self-explanatory quote gives a clear idea about some of the reasons behind a low level of technological integration, but also more generally behind the adoption of digital technology. It doesn't make sense for our entrepreneurs to adopt communication technologies if they cannot use them to communicate with their suppliers. In this case, the suppliers are local dairy farms.

Managerial cognition

Given the lack of skills and knowledge among owners or managers of agri-food SMEs in the field of digital technologies, it is important to assure a constant flow of knowledge. The process of digital technology adoption also entails an organizational change and therefore owners or managers are forced to face several complexities. The context in which these enterprises are operating is nowadays characterized by a constant level of change. To be flexible and successfully capable of innovating, promoting a learning culture is fundamental.

In a climate where such companies operate, continuous change is the norm. So, the capacity to innovate in response to change is pivotal for the organizational development and might also be achieved through the promotion of a "learning culture" (Gilbert and Cordey-Hayes, 1996). In line with this concept, we observed that some HTI firms' managers understood the importance of the learning culture and created an environment where everyone in the company is willing to learn:

> My employees are curious, they are passionate about technology, they subscribe to industry magazines, they keep me informed, reporting to me things that I didn't even know myself [...]. Sometimes it's me who has to say stop to them; I tell them that I have already done enough, I want to stay like this for a bit ... Instead, they bring me to check out new technologies and try them [...]. You need to have knowledge about what you buy.
> (Manager of Company 8)

The need for a constant flow of knowledge is underlined in this quote. The manager perfectly understood the need for constant knowledge, and in response to this, he was able to create a learning environment where each employee is active in this sense. Within this company, the process of gaining knowledge takes place automatically through both the manager and the employees. It is no coincidence that this firm massively relies on digital technologies for its daily operations. However, in our sample, we also observed a similar cognition in another HTI firms' manager. Here is a quote by the manager of Company 13, who is referring to technology and technology providers:

> even if someone has outdated knowledge, you always need to ask them (*ask technology providers used as a source of knowledge*), you need knowledge on a constant base; you at least need to have a look to see what is going on or to refresh your knowledge in order to understand what you have lost because of the frequency of the business's activity.
>
> (Emphasis is mine)

Instead, among the LTI firms' managers interviewed, we did not observe the same effort deployed by the HTI firms' managers to guarantee a constant flow of knowledge. The absence of this cognition is probably a consequence of the lack of importance given to technological innovation. The fact that they don't understand how important digital technologies are to sustain competitive advantage leads them not to invest time and resources in encouraging a constant flow of knowledge. An emblematic and self-explanatory example of this attitude is provided by the manager of Company 19, who stated the following:

> I don't stimulate knowledge acquisition, in the sense that I gain knowledge casually. I talk to my colleagues or with other people who tell you if they tried this or that technology. I am not proactive in the process of acquiring this information.

Managerial behavior

From a behavioral point of view, we observed that the HTI enterprises differ from the LTIs because the former sustains these types of behaviors while the latter doesn't:

> Proactive learning behavior in the acquisition of new and relevant knowledge (B1);

Constant attention to emerging challenges and novelties (B2).
The connection with the aforementioned belief is clear. If the core belief of a firm is based on technological innovation, new knowledge about digital technology should be constantly added and the emerging challenges brought on by globalization and technological novelties should constantly be researched. Specifically, in B1, proactive behavior is defined as all the actions undertaken by the manager to acquire knowledge about the best sector-specific technologies available on the market and how to include these technologies within the firm's

business model, anticipating even the early majority of adopters (Lumpkin and Dess 1996). In this regard, the manager of Company 7 said:

> The first factor for success is the curiosity of the owner/entrepreneur because curiosity opens you up to novelties and the use of a network. Network is a facilitator.

Proactive behavior is the consequence of the curiosity of an entrepreneur who is seeking new knowledge (Crant, 2000). The search for new knowledge creates the necessity of being open, considering and listening to multiple external sources of expertise, experiences, and opinions coming from the same sector or other sectors. As far as B2 is concerned, the manager of Company 6 said:

> You are active in the search for information, doing it every day because you always find new stimuli.

From this quote we understand that "constant attention" means everyday attention, underlining the manager's vast commitment to seeking out challenges and novelties. Within the boundaries of the business of Company 6, one of the emerging challenges that were mentioned in the interview is the low price of milk that is being paid to farmers. To tackle this challenge, the manager adopted what in B2 is called a "novelty," that is, the voluntary milking system he adopted on his farm.

On the contrary, the managers of the LTI enterprises did not show the behaviors present in B1 and B2, typical features of HTI enterprises. LTI appeared to be much less proactive in acquiring relevant knowledge and less concerned about emerging challenges and novelties. For instance, the manager of Company 4 said:

> I don't compensate for the lack of knowledge. I try to do what I can. I don't usually look for knowledge. This is also because if I don't have the right skills, I prefer to take it easy and take my time.

Managerial practices

Usually, the decision to adopt a digital technology comes after a relatively careful evaluation of certain criteria. This evaluation is undergone by the managers of both HTI and LTI enterprises, though the groups differ in terms of the quality and quantity of criteria considered. On the one hand, the managers of LTI enterprises have been shown to make relatively poor judgments, mostly based on two factors:

> *The real need for digital technology;*
> *The cost of the investment.*

On the other hand, the mangers of HTI enterprises usually make a much deeper analysis when adopting new technologies, including the criteria stated above plus additional ones. Here are some of the additional criteria that were mentioned during their interviews:

> Creation of a business plan, including the SWOT analysis, for investing in technology.
>
> (Company 6 and 10)

> Evaluation of the effort in implementing technology in terms of the time needed to switch from the old technology to the new one.
>
> (Company 10)

> Assessment of the cost and economic return on investment, advantages in the logistics, and optimization of production processes.
>
> (Company 7)

> Evaluation of employee safety and economic return on investment.
>
> (Company 8)

> Evaluation of the cost and economic return on investment, as well as savings in terms of employee deployment.
>
> (Company 9)

Once the digital technology has been adopted, we observed that the managers of HTI enterprises usually evaluate the performances of the business using formal benchmarking activities. Of course, given the fact that their strategy is mainly based on adopting digital technology, it is not reliable to only count on subjective evaluations of firm performance, such as in the case of LTI firms' managers. The difference between the two groups is shown here through different performance evaluations. High-tech managers said:

> We use benchmark indicators that we find available on the market. When we adopt new technology, we refer to our historical data as well.
>
> (Company 7)

On the other hand, low-tech managers said:

> I evaluate the performance by eye. It is useless to talk about percentages when the bases are so paltry.
>
> (Company 5)

The lack of rigor in practices is also visible between the groups when we observe the frequency with which the strategic choices are checked. The pattern of the interview answers is also similar in this regard, showing a systematic approach in

HTI enterprises. The managers of HTI enterprises check their strategic choices constantly:

> We check our strategic choices every day [...].
>
> (Manager of Company 6)

The managers of LTI enterprises check them more sporadically:

> I never check our strategic choices.
>
> (Manager of Company 2)

In this regard, it is interesting to notice that the difference between the two groups is not only related to the frequency with which strategic choices are revised but also the presence or absence of a real strategy. It is common in agri-food SMEs to not have a real strategy (Costa and Jongen, 2006) because the business rhythm is dictated by the rate of everyday activities. The lack of a clear strategy can also be seen in the degree of the vagueness of the answers by some low-tech managers:

> We rethink our strategic choices gradually. Maybe this year you innovate for something and next year for something else.
>
> (Manager of Company 3)

Another difference between the two groups of SMEs emerged regarding their exposure to relevant information. The managers of HTI enterprises usually vaunt higher exposure to important information about digital technology. As the manager of Company 6 stated:

> I find ways to collect information through other people or magazines, or through other peers, to evaluate certain factors, which then allows me to make a managerial decision.

In the interview with the manager of Company 2, on the other hand, she had never gone to expositions or events related to digital technology, and that she is only in contact with peers, owners of businesses similar to hers. These peers are also entrepreneurs whose businesses are not based on digital technology, therefore they do not represent relevant sources of information about technology.

Furthermore, we observed that indistinctively among the HTI and LTI groups, feedback from customers is used as an incentive for adopting and implementing digital technology. The feedback is collected, in most cases, orally by the figures who are directly in contact with the end customers. Below is an emblematic quote, in which the manager of Company 15 (LTI) actually decided to adopt a business software based on feedback provided by a specific group of customers, ethical purchasing groups (GASes). These groups' request is very specific because they buy products as one big collective order. This order is comprised of several small orders for each household that is part of the group.

Therefore, the firm has to manage many different packages at once and consequently has to be very flexible.

> […] we received a big impetus from the GASes because we were not able to process their orders anymore by only using Excel […]. Not only did they give encourage us, but they also helped us in the process of adopting the enterprise software because some of them are very skilled in computer science.

Conclusion

In this study, we report how managers of agri-food SMEs enable the successful adoption and usage of digital technologies. With a comparison between the two groups of cases, we emphasize the managerial characteristics that are related to a higher level of integration of digital technology within the firm. Overall, we can say that all the differences between HTI and LTI enterprises can be summarized in the presence or absence of certain rigor applied to business practices. For HTI enterprises, we observe clarity, a systematic approach, and logics set on adopting digital technology. As for the other managers, those of LTI enterprises, we observe the absence of these practices and a major degree of disorientation toward the process of technological innovation.

References

Annosi, M. C., Brunetta, F., Monti, A., & Nati, F. (2019). Is the trend your friend? An analysis of technology 4.0 investment decisions in agricultural SMEs. *Computers in Industry*, 109, 59–71.

Bantel, K. A., & Jackson, S. E. (1989). Top management and innovations in banking: does the composition of the top team make a difference? *Strategic Management Journal*, 10(S1), 107–124.

Baourakis, G., Kourgiantakis, M., & Migdalas, A. (2002). The impact of e-commerce on agro-food marketing: the case of agricultural cooperatives, firms, and consumers in Crete. *British Food Journal*, 104(8), 580–590.

Berergon, F., & Raymond, L. (1992). Planning of information systems to gain a competitive advantage. *Journal of Small Business Management*, 30(1), 21–26.

Bertrand, M., & Schoar, A. (2003). Managing with style: the effect of managers on firm policies. *Quarterly Journal of Economics*, 118(4), 1169–1208.

Bridge, S., O'Neill, K., & Cromie, S. (2003). *Understanding Enterprise, Entrepreneurship and Small Business*, 2nd ed. Palgrave Macmillan, Basingstoke and New York, NY.

Charmaz, K., & Belgrave, L. (2012). Qualitative interviewing and grounded theory analysis. *The SAGE Handbook of Interview Research: The Complexity of the Craft*, 2, 347–365.

Corbin, J. M., & Strauss, A. (1990). Grounded theory research: procedures, canons, and evaluative criteria. *Qualitative Sociology*, 13(1), 3–21.

Costa, A. I., & Jongen, W. M. F. (2006). New insights into consumer-led food product development. *Trends in Food Science & Technology*, 17(8), 457–465.

Crant, J. M. (2000). Proactive behavior in organizations. *Journal of Management*, 26(3), 435–462.

Damanpour, F., & Schneider, M. (2008). Characteristics of innovation and innovation adoption in public organizations: assessing the role of managers. *Journal of Public Administration Research and Theory*, 19(3), 495–522.

Davis, F. D. (1989). Perceived usefulness, perceived ease of use, and user acceptance of information technology. *MIS Quarterly*, 319–340.

Durst, S., & Runar Edvardsson, I. (2012). Knowledge management in SMEs: a literature review. *Journal of Knowledge Management*, 16(6), 879–903.

Eisenhardt, K. M., & Graebner, M. E. (2007). Theory building from cases: Opportunities and challenges. *Academy of Management Journal*, 50(1), 25–32.

Geletkanycz, M. A., & Hambrick, D. C. (1997). The external ties of top executives: Implications for strategic choice and performance. *Administrative Science Quarterly*, 654–681.

Gilbert, M., & Cordey-Hayes, M. (1996). Understanding the process of knowledge transfer to achieve successful technological innovation. *Technovation*, 16(6), 301–312.

Gioia, D. A., Corley, K. G., & Hamilton, A. L. (2013). Seeking qualitative rigor in inductive research: Notes on the Gioia methodology. *Organizational Research Methods*, 16(1), 15–31.

Giotopoulos, I., Kontolaimou, A., Korra, E., & Tsakanikas, A. (2017). What drives ICT adoption by SMEs? Evidence from a large-scale survey in Greece. *Journal of Business Research*, 81, 60–69.

Hambrick, D. C., & Mason, P. A. (1984). Upper echelons: the organization as a reflection of its top managers. *Academy of Management Review*, 9, 193–206.

Jarvenpaa, S. L., & Ives, B. (1991). Executive involvement and participation in the management of information technology. *MIS Quarterly*, 15(2), 205–227.

Jeyaraj, A., Rottman, J. W., & Lacity, M. C. (2006). A review of the predictors, linkages, and biases in IT innovation adoption research. *Journal of Information Technology*, 21(1), 1–23.

Kane, G. C., Palmer, D., Phillips, A. N., Kiron, D., & Buckley, N. (2015). Strategy, not technology, drives digital transformation. MIT. *Sloan Management Review and Deloitte University Press*, 14, 1–25.

Kuan, K. K., & Chau, P. Y. (2001). A perception-based model for EDI adoption in small businesses using a technology – organization – environment framework. *Information & Management*, 38(8), 507–521.

Lefebvre, L. A., Mason, R., & Lefebvre, E. (1997). The influence prism in SMEs: the power of CEOs' perceptions on technology policy and its organizational impacts. *Management Science*, 43(6), 856–878.

Li, L., Su, F., Zhang, W., & Mao, J. Y. (2018). Digital transformation by SME entrepreneurs: a capability perspective. *Information Systems Journal*, 28(6), 1129–1157.

Lin, T. C., Ku, Y. C., & Huang, Y. S. (2014). Exploring top managers' innovative IT (IIT) championing behavior: integrating the personal and technical contexts. *Information & Management*, 51(1), 1–12.

Low, C., Chen, Y., & Wu, M. (2011). Understanding the determinants of cloud computing adoption. *Industrial Management & Data Systems*, 111(7), 1006–1023.

Lumpkin, G. T., & Dess, G. G. (1996). Clarifying the entrepreneurial orientation construct and linking it to performance. *Academy of Management Review*, 21(1), 135–172.

Mehrtens, J., Cragg, P. B., & Mills, A. M. (2001). A model of Internet adoption by SMEs. *Information & Management*, 39(3), 165–176.

Midavaine, J., Dolfsma, W., & Aalbers, R. (2016). Board diversity and R&D investment. *Management Decision*, 54(3), 558–569.

Nguyen, T. H. (2009). Information technology adoption in SMEs: an integrated framework. *International Journal of Entrepreneurial Behavior & Research*, 15(2), 162–186.

Pickernell, D. G., Christie, M. J., Rowe, P. A., Thomas, B. C., Putterill, L. G., & Lynn Griffiths, J. (2004). Farmers' markets in Wales: making the 'Net work? *British Food Journal*, 106(3), 194–210.

Premkumar, G. (2003). A meta-analysis of research on information technology implementation in small business. *Journal of Organizational Computing and Electronic Commerce*, 13(2), 91–121.

Premkumar, G., & Roberts, M. (1999). Adoption of new information technologies in rural small businesses. *Omega*, 27(4), 467–484.

Rizzoni, A. (1991). Technological innovation and small firms: a taxonomy, *International Small Business Journal*, 9(3), 31–42.

Shah Alam, S. (2009). Adoption of internet in Malaysian SMEs. *Journal of Small Business and Enterprise Development*, 16(2), 240–255.

Teece, D. J. (2017). Towards a capability theory of (innovating) firms: implications for management and policy. *Cambridge Journal of Economics*, 41(3), 693–720.

Tikkanen, H., Lamberg, J. A., Parvinen, P., & Kallunki, J. P. (2005). Managerial cognition, action and the business model of the firm. *Management Decision*, 43(6), 789–809.

Walsh, J. P. (1988). Selectivity and selective perception: an investigation of managers' belief structures and information processing. *Academy of Management Journal*, 31(4), 873–896.

Warner, K. S., & Wäger, M. (2019). Building dynamic capabilities for digital transformation: An ongoing process of strategic renewal. *Long Range Planning*, 52(3), 326–349.

Winston, E., & Dologite, D. (2002). How does attitude impact IT implementation: a study of small business owners. *Journal of End User Computing*, 14(2), 16–29.

Yin, R. (2003). *Case Study Methodology: Applied Social Research Methods Vol. 5*. London: Sage Publications. Thousand Oaks, CA, USA.

Young, G. J., Charns, M. P., & Shortell, S. M. (2001). Top manager and network effects on the adoption of innovative management practices: a study of TQM in a public hospital system. *Strategic Management Journal*, 22(10), 935–951.

4 The use of digital technologies in agri-food
Evidences from the rural sector

Maria Carmela Annosi and Federica Brunetta

Synopsis

In 1984, Company A was founded with the specific aim of rationalizing agricultural activities on the lands belonging to its founders' family. They started with 40 hectares, and over the years they acquired 400 additional hectares of arable land. Integration of new advanced technologies to assist production was central to their mission and was a means for them to grow and compete. Let's discover how they organized the firm for the adoption and integration of new digital technology to become competitive.

Company B, on the other hand, was founded in 1939. In the beginning, the company based its business on just 30 cows. At that time, the milk utilized was cooled at four degrees and corked in bottles by hand before being distributed throughout the surrounding area. In 1966, the second generation founded another company with a new name, becoming one of first companies devoted to processing milk. The third generation decided to go back to the original production of fresh and genuine products. This decision led the third generation to launch a new dairy company strongly connected to the land, seen as an answer to those searching for tastes of the past. Let's see how the third generation of founders, not being digital natives, was able to cope with the difficulty of integrating and using new technology.

> **Case A: the successful case of technology use in Company A**
>
> *We are an open firm. We have always been open. Our company was founded by my father, who died in a car accident when he was 48. At that time, I was 11 and my brother was 18. We were able to make a living with the support of my mother, who had assisted my father in our business, so she knew exactly what she was doing. At that time, this was enough to survive. Having a bit of experience could be enough, but it is no longer that way since our business has grown increasingly complex. What we have always had, however, is a curiosity to know other people and an ability to identify who counts in our fields and who can lead, or at least who we can learn from, and we found many of them.*

Founder and manager of Company A

Company A was founded in 1984 with the objective of rationalizing agricultural activities on the lands belonging to its founders' family. The family vaunted immense experience in the agricultural sector with 50 years of successful activity across three generations. Their main business was initially the cultivation of cereals, excluding rice. Then, in 1984, the last generation decided to renew and enlarge the scope of the firm's operations, and to manage their B2C business (tied to the cultivation of cereals) together with a B2B business offering consultancy services to other farms. They called their new business "by agricultural entrepreneurs for agricultural entrepreneurs."

They started with 40 hectares of arable land, but due to the success they achieved thanks to their strategic choices and the successful use of the newly adopted technologies, over the years, the renewed company grew to extend its activities to over 400 hectares, achieving a sales revenue of 1.3 million euro.

The adoption of precision farming, meaning the use of tractors and operators that are able to modify their operations to cope with the variabilities they encounter, allowed them to implement traceability in their cereal productions. Within the precision farming's possibilities, they used DVI images to build prescription maps and everything needed for cultivating cereals. The cereals produced could be optimally conserved due to the use of silobags, reducing the concentration of oxygen and increasing the concentration of carbondioxide. They also succeed in creating the first electricity-generated, 250 kWp solar greenhouses, with the aim of achieving greater social and economic impact. They also adopted software programs that allowed them to continuously observe costs for each plot of land and each cultivation type. Through this close monitoring, they were able to normalize costs for 1 kg of cereals. Each employee was also equipped with a tablet so that each intervention made could be registered and saved in the cloud.

Since its foundation, the company has had a clear vision, which its founders advertised quite extensively through different media: "Innovation means working everyday with the ambition to improve." The company also had a clearly-communicated mission containing a few relevant social objectives: "Agriculture is our mission. We consider the development of agriculture of primary importance within the economy to the benefit of everyone and the environment."

Founders of the company also defined the five main pillars feeding their interests and justifying their daily operational activities while contributing to the overall mission and vision of the company: Productivity, Sustainability, Competitivity, Smart, and Resilience. They explained their dedication to these five pillars with the following explanation.

- *Productivity*: the increasing request of agri-food products due to the global population growth creates an urgent need to use as few natural resources as possible and as little land as possible.

- *Sustainability*: constant attention to the environment, shared goods, and reducing the use of chemical addictions.
- *Competitivity*: capacity to remain on the market, rigorous monitoring of costs, and innovation in the production processes.
- *Smart*: use of information and more advanced electronic devices to optimize our interventions and make all our products traceable (precision farming).
- *Resilience:* capacity to face and overcome negative events, such as climate change, and the difficulty of the markets in creating an opportunity from problems.

The founders of Company A had their own network of contacts outside the traditional trade associations. They indeed did not trust the traditional trade associations anymore due to its past attempts to do business through the services offered, deviating from their original mission of support. Therefore, the founders preferred to attend an informal network of entrepreneurs, where they could find greater openness to their ideas and needs. They met members of this network on a monthly basis. During these meeting, they were able to discuss different themes specifically linked to their activities, the technology to adopt, the market difficulties they faced, and the emerging business opportunities they observed.

The meetings among this informal network were regularly attended by representatives of about 30 firms, but the actual number of the network's members was much higher. For the entrepreneurs, attending those meetings was not always easy because of various urgencies or constraints, but many of them were also able to connect via video conference or Skype.

Within this network, they established preferred relationships, including universities. With some of them, they even established long-term partnerships based on their common interests. Thanks to these contacts, the connection was strengthened by weekly calls and quick, on-demand comparisons regarding specific problems when they emerged.

The founders defined their firm as open, meaning a firm without boundaries, with each one of them having increasingly complex working relationships and able to gain rapid access to knowledgeable people and establish multiple collaborations. Their firm was founded by their father, who died at the age of 48 when his children were 11 and 18 years old. Their mother taught them the rules of their business since she had assisted their father in its development. She knew what she was doing and understood the basic rules of the market. At that time, this was enough to survive. But later on, the founders realized that the reality of their business environment had become much more complex and that what the mother taught them was no longer enough. Nevertheless, since its inception, the company's founders have vaunted a curiosity to know other people and an ability to identify those who, according to them, could teach them more or who were leading thinkers. They started travelling a lot to meet important entrepreneurs and very good technicians. They started very fruitful collaborations with a few English agronomists that opened their horizons and changed their approach to the work and the cultivation of cereals. With this, they maintained a continuous exchange of information and ideas. Many other foreign contacts visited them as they were always available to talk and share their business ideas with new and interesting people. There was a sort of rule of reciprocal help among their peers, as they had always found availability when it was their turn to make a visit. Quite recently, given the increasing relevance of their brand, they have started attracting even more visitors than in the past.

Regarding the collaboration with technological partners, the founders preferred having business with ones that they already knew and with whom they already had previous collaborations. This principle was certainly followed for the purchase of proper machinery for precision farming. For less vital services, such as purchasing their drones, they preferred to follow the best ratio quality price. In both the circumstances, they also purchased the support and training for their employees.

Even though they continuously searched for new and interesting contacts for their business, the founders of Company A felt the need to have greater continuity in the information provided by their technology suppliers, which instead only provided information for the specific machines sold or to sell. The founders or managers of Company A were indeed aware of the volatility of the technology, especially in the context of precision farming. They were also aware of how difficult it had been for other competitors and farmers in their geographical region to keep up the technological evolution.

The relationships between founders were very good, with just a few conflicts. They clearly understood which principles to use to properly regulate their collaborations. Transparency, trust, esteem, and ability to delegate were considered crucial. They also believed that none of them could presume to be demanding. They believed in the principle of flexibility to avoid creating a bottleneck or structural weaknesses due to dependence. Every company decision was made together with the other founders and every relevant piece of information was shared. This was also to foster an interchangeability of roles within the company. Once the decision was made, all the founders had to execute it by aligning their practices.

Internally, the information about a new technology to adopt could be easily spread given the limited number of employees (there were only two others in addition to the three founders). Formal events with technology suppliers were also organized to favor an exchange of relevant technological information between the suppliers and employees. During these events, there were discussions about the implications of using the technologies, the firms' performance, and the way employees worked.

Employees were included in the decision to adopt a specific technology. They were also encouraged to provide their ideas on how to improve certain application.

However, the founders of Company A only hired employees with a high education level, as this meant they could use and apply the new technologies.

Given the limited number of employees and the constant interaction with them, the founders had direct and informal contact with them, using less written communication to update them on the latest news or events, while the employees constantly informed the founders of any deviations observed in the fields. Everyone has a tablet and company cell phone with which they could constantly communicate and provide real-time data.

In order to increase their focus on the more important things to do, the founders organized informal coordination meetings with employees every morning before leaving for the fields. In these meetings, employees were informed of updated plans and relevant information for carrying out their tasks and provided quick feedback about what happened the day before.

There was a reciprocal bottom-up/top down flow of information between the founders and the employees, who could be both passive and active in giving relevant information related to managing the fields.

Individual employee performance was also constantly monitored, since employees had to register all their operations in the cloud, reporting the time needed for each executed operation and its results. Each employee had a tablet through which he/she could easily upload all the necessary information onto the company cloud. For instance, they had to declare if they had used the tractor and for how long and if they had added chemicals and/or other fertilizer to the soil. This was important for the traceability of the production process. All their products had a QR code that reported all the traceable information about their production and the specific piece of land they belonged to.

Employees were continuously formed beyond the institutional obligations. Training programs were established and planned every time there was a new machine to use. The founders asked their employees to read and study all the manuals before attempting to use the new machines. This meant that there was a strong focus on training given the complexity of their new technological instruments.

Founders paid constant attention to new technologies. They were always searching for new possibilities. The adoption of new technologies had to follow an incremental process by looking at what was available on the markets. One of their objectives was to advance the application of precision farming in order to have more precise estimations and data to rely on.

The decision to adopt a new technology went through an in-depth analysis of the economic advantages the firm might benefit from. As such, the value of each technology was assessed through the perception of the potential impact it could have on the entire organization.

Every strategic decision for the firm related to the adoption of any technology was also internally reviewed and studied by constantly monitoring its related performance.

The impact of the technology on the company's performances was evaluated by monitoring the historical data they kept on specific indicators. This provided objective data with which they could decide the relevance of a certain technology for the firm.

Regarding the managerial style driving their business, the founders were aware that their curiosity to learn was the true basis that could justify their firm's success. They believed that searching for novelties and being part of a network could help a lot. They believed their presence in the network was a successful element for their business, as their social network was considered a means to share information, to learn new things, and to do things fester. They also believed in the need to create collaborations and to achieve things alongside with others in order to accomplish their goals more easily. Compared with the strategy adopted by other farmers, who, due to their isolation, tended to think they were operating properly, the founders of Company A were always looking for other solutions to compare with and learn from. Additionally, their presence in their social network had helped them to overcome the negative aspects of the markets, dominating the complexity of the new technology.

They also strongly believed that the problems encountered by local farmers were due to the absence of entrepreneurial skills in many managers or owners. They felt that adopting new technologies was crucial to a firm's importance.

They actively stimulated their knowledge through the exchange of information among the network they belonged to and the dyadic relationship they could establish with universities and other partners.

They also referred to the lack of closeness with academia and field experts, who were somewhat distant from the real problems they faced, meaning there was a need to have them on board.

Case B: a story of technology change, the case of Company B

The process of automatic milking enacts controls over the somatic cells (CS) of milk, the monitoring systems for cows detects their movement, and robots evaluate the quality of the milk, but all these devices do not reveal how a cow really feels. The computer does not reveal if a cow is not feeling well, if it may have some difficulty in walking, or be slow and have drooped ears. There are a series of elements that only the owner of a cow can recognize as symptoms of problems. It is important to know the company thoroughly, from the most minute detail to the last liter of milk produced, if one wants to be able to apply technology and produce something that can really help farmers, given the specific peculiarities of their farms.

Founder and manager of Company B

In 1939, the initial founder of company Y established the first farm with just 30 cows. In those years, the milk was cooled at 4 degrees and corked in bottles manually. In 1966, the founder's eldest son launched a factory for processing milk. The factory was among the first in Italy to be devoted to processing milk.

The third generation of owners decided to go back to the original production of fresh and genuine products.

The third generation's owners demonstrated a strong willingness to obtain very high-quality products. In 2018, Company B vaunted 350 animals, including cows and buffaloes, and produced more than 30 different kinds of cheese as well as fresh milk, mozzarella, and yoghurt. In addition to the farm, the company had a restaurant where they served different types of bread, biscuits, and pizza, all made using their own cereals and legumes. The third-generation founders also opened a shop that operates even on holidays. Near the farm, they also established a picnic area where visitors could also eat outdoors. Many schools visited the farm, sometimes organizing small experiments.

Company B, indeed, implemented the concept of an agri-food supply chain where only a few intermediaries are used between producer and consumer. They also placed great emphasis on other aspects, such as quality, origin, and the "naturality" of agri-food production. To avoid the influence of intermediaries, they used products from their land to feed their animals. This allowed the founders to exercise better control over prices, which could be determined autonomously. This approach meant the farmers could also regain control over decisions about what to produce, escaping the vicious circle typical of traditional markets. This also meant they could avoid being pressed by their suppliers and by the wholesalers they sold their products to, losing their decision-making autonomy.

Company B did not make large use of digital technology. The most advanced digital device used was a long-distance, terminal-based system that provided a complete reproduction monitoring solution for their cows. This system was felt to be particularly useful for the type of cows they raised, as they were not as fertile. The digital system they adopted gave them a good degree of support in predicting optimal moments for the cows to reproduce. It entailed multifunctional neck tags with a proprietary movement sensor and a rumination recorder. To also use alternative sources of energy, they had a photovoltaic plant installed and an agricultural biogas plant, helping them earn additional profits. Furthermore, the founders adopted digital control over the devices used during the production phases, so they could remotely intervene in case any of the parameters deviated.

The founders were on the board of directors of a national trade association, with the aim of representing agriculture at the national level and protecting the interests of farmers Italy-wide. However, the discussion of which technologies to adopt and in which context was never touched upon. Nevertheless, the founders vaunted an informal network of peers operating in the same field that they could interact with in order to trigger discussions about the technology they adopted. This informal network diminished over the years, however. In 2018, just five farmers were involved, and the main discussions that emerged were related to renewable energy and other disposal systems to carry out monitoring and checking up on the animals. The founders had monthly meetings with the other members of this informal network.

They tried to use the network on demand needs and developments, just to have easier access to the bulk of the consolidated experience of others in the network. The use of technology was never considered a priority. This also justified their tendency to skip informational events about new technologies. In addition to having support available – also considering the geographical distance of suppliers from where they were located – they did not use any other specific system or criteria to evaluate the risk of introducing new suppliers to their business. Sometimes they used their instinct to decide whether to trust them or not. They usually searched for technological partners by starting from a specific need/development, before deciding who could offer the right answer to their question but under conditions that could inspire trust in them. For them, trust was a matter of feelings and impression; it was not something that could easily be explained.

The firm demonstrated a clear hierarchical system, with the main owner appointed to coming up with proposals for utilizing resources and selecting the decisions to be implemented. Carrying out the ratified decisions and managing the risks were the responsibility of the other two founders. Being more on the operative side, the other two founders were also involved in measuring the effects of the decisions made. Essentially, Company B created a separation of decision and risk-bearing functions, specializing in and benefitting from this kind of work organization.

The primary motivation for adopting new technologies was the claimed and projected benefits these technologies could bring to the company. The choice to adopt or reject an innovative decision passed mainly through the hands of the firm's owner, though only after consulting with the other two owners, who were called in to provide their technical expertise about their understanding of how the technological innovation could be used to overcome existing performances and difficulties. Within the company, no formal event was specifically organized to share information or to discuss the firm's performance. Owners were used to informal talks mainly driven by the need to coordinate internal activities on the farm. In the context of this company, the communication culture was usually verbal, informal, and "in the hallway." Therefore, knowledge tended to be passed on without any associated records or documentation because of their informal nature.

The same happened when coordinating the activities, which, with the rest of employees, occurred through instant messaging. Through these text messages, the owners allocated tasks to employees and collected information about the progress of their activities.

Regarding organizational learning, it occurred at all levels and functions in the firm, with the owner or manager of the company eager to be informed about what his employees learned during the day's operations. But he also collected information about new technology, with the duty of distributing this information throughout the firm in order to discuss the implications derived from adopting new solutions for the farm. More in general, the entire firm relied more on informal rather than formal learning programs due to their lack of resources, with learning more incidental and reactive.

The adoption of a new technology was never planned in advance. The adoption process would start from receiving useful input. The founders had only planned for what they were currently using, such as the system used for the cows. Three elements should be concurrent when organizing the adoption of a technology: the availability of resources to invest in, the presence of a latent need, and meeting with some technology suppliers and inspiring trust.

Innovation was not planned strategically and, as such, the decision to adopt a technology was also never reviewed and reflected upon.

When the use of technology and the benefit they could get from the technology were clear, then the firm would plan. They did that in one case, but in that situation, the local university they collaborated with had some sort of internal problem and the project was unfortunately not carried out.

For the firm, it was also not clear and easy to determine the impact of a technology on the business's performance. The preventive evaluation of the impact of a technology required them to have a long period of observation across different seasons. This was the case for the device used for the cows. They had to fully test the digital system for six months before being able to assess its efficacy and the impact it would have on their business.

One of the factors considered crucial for the successful diffusion of the technology within their company was its ease of use and whether the technology could work as expected. The expectations they had for the technology were mostly focused on the need for it to be easy to use and on an evident practical implication. For instance, the company's founders underlined their desire to have more automation and remote controls within their field so as to reduce the amount of resources utilized.

The founders also underlined the need for the designer of the technology to truly understand the practical context of its application and to use the technology, so they could understand its consequences. They tended to prefer technologies that were more sophisticated but that were simple to use, and which had a profound effect on the company's organizational structure and their decision-making abilities. This would require deep knowledge about the context of the application, which is not always easy to get.

For instance, for their biogas plant, the founders were using a very easy app through which they could control and manage the entire system. Using technology should be seen as easy and instinctive. In the case of the digital device applied to the cows, they could not rely entirely on that system to fully know how the animal was, which required them to carry out additional, constant checks on the animals.

But the company was not active in the search of new solutions. Access to relevant information occurred through random interactions and exchanges of information among friends and peers. They were not even interested in attending informational events.

Part III

The emerging promises of Internet of Things (IOT) and digital ecosystems in agri-food

5 The power of business ecosystems in the agri-food sector
An interview with Sjaak Wolfert

Maria Carmela Annosi and Teun Gilissen

Background

How do you manage the coordination of business ecosystems in the agri-food sector?

Sjaak Wolfert is a senior scientist at Wageningen University and Research. He studied the socio-economic aspects of data sharing, business modelling, and governance issues within business ecosystems. He is the scientific coordinator of (inter)national projects such as the EU-project Internet of Food and Farm (IoF2020), SmartAgriHubs, and DATA-FAIR. He is affiliated with the Information Technology Group at Wageningen University and was president of the European Federation of ICT in Agriculture (EFITA). Sjaak is a visionary, challenged by complex problems that require a science-based approach in which organizational and technical aspects need to be combined.

Maria Carmela Annosi and Teun Gilissen visited Sjaak at his office at the fourth floor of the Atlas building at the Wageningen campus.

In this interview with Maria Carmela Annosi, author of this book, and Teun Gilissen, research assistant at Wageningen University, Sjaak Wolfert described how his institution could manage the development and coordination of large business ecosystems in the agri-food sector and how SMEs are using digital technology and their presence within the business ecosystem to blur the traditional boundaries of their market through new business models. In so doing, SMEs are evolving in a direction that Sjaak Wolfert calls disruptive as even farmers are now becoming aware of the relevance of their data and the opportunity to open new business around it. Sjaak Wolfert also mentioned the emerging governance problems in coordinating these business ecosystems. He referred to the difficulty in building a public sphere of communication of ideas and projects within the ecosystems given the enormous difference in knowledge backgrounds among the partners involved. Through his insights, it also comes clear that the process of sharing data, ideas, and resources within the ecosystem has shifted the debate among ecosystem partners to a local debate on who has to own the data and how to offer access to the data. This prompts the emergence of ad hoc forms of local governance within the ecosystems. Wageningen Economic Research emerges as a mediator between farmers, technological providers, and civil society, working to ensure that the balance between stability and the change introduced is always maintained in the conduct of the EU projects. The direct model of the EU projects' structures also provided benefits in keeping in line the network within the ecosystem, but once the EU projects are over, and the mediation provided by Wageningen's neutral coordination ends, then farmers and other partners can fail to fulfil the demand of the interactions established during the projects. In addition, if the channel of communication between two or more parties is blocked, the whole system of interdependencies comes to a stalemate. But as Sjaak Wolfert recognizes, inside the business ecosystem there is a variety of social and private international, conational, and political institutions possibly in the position to enact ad hoc forms of global governance, allowing all the partners to interact in a non-disruptive manner.

Interview

Maria: Could you please introduce yourself and tell us about your role and related work experience with EU projects in the area of digitalization in the agri-food sector?
My name is Sjaak Wolfert. I work at Wageningen Economic Research, which is one of the research institutes at Wageningen University and Research (WUR), as a strategic senior scientist. My research interests are in the area of digital innovation in the agri-food business. I am mainly working on the coordination of large European projects, which usually rely on public–private partnerships. These projects are executed with the help of public institutions such as WUR, but also other universities and research institutes in other countries, as well as private companies operating mainly within the agri-food and ICT sectors. These projects usually include both large and small companies. However, the majority of companies involved are SMEs. We have sometimes also worked on special accelerator programs, in which start-ups were involved as well. My role in these projects is the scientific coordination. Therefore, I am looking at the content of these projects by working on project proposals, and at the progress reported by the involved partners. I am also the author of several scientific publications that use data and experience from these projects.

Teun: Could you tell us a bit more about how these projects are organized and how they are connected to each other? What do they have in common?
From 2010 onwards, we have been involved in a series of EU projects. We started with Future Internet (FI) programs funded by the European Commission (EC), with the purpose of building a common platform for new internet developments like the Internet of Things (IoT), big data, and later on, artificial intelligence, blockchain, and so forth. These are generic technologies that every industry is challenged with. Every sector – energy, automotive, health, but also agriculture and food – is looking into this. But how can they use this? The Future Internet Public–Private Partnership projects are aimed at accelerating the development

and adoption of Future Internet technologies in Europe and advancing the European market for smart infrastructures.

They started adopting a use case approach, meaning that every sector had to try out several concrete applications, using this common platform. Large telco companies developed the common platform. The use case approach is still applied and is the heart of our approach for all the projects that we lead. The underlying assumption of having use cases is that the knowledge generated will be modular and re-usable across domain or use case-specific boundaries. Each use case analyzes the generic developments of the technology as well as aspects like organizational challenges, impact on business modeling, and governance.

Maria: *So how do you organize your research work in each use case with these projects?*
We started focusing more on aspects related to data sharing. Then our interest became broader and we included the analysis of legal and regulatory effects of the establishment of business ecosystems. Just recently, we further broadened our area of interests, looking at the ethics part of the collaboration among actors. Basically, these use cases are the unit of analysis of our project. Each use case is organized through an ecosystem of business actors which are connected to each other, helping them to spread digital technologies in the European agri-food sector.

We provide each use case with technical and organizational support. At the same time, we try to devote time to the development of the entire ecosystem. These ecosystems involve farmers that we try to attract by organizing typical field days, in which a lot of farmers come and look at new developments. This is our main approach that we have applied in all the projects to date.

Teun: *You already touched upon how the ecosystem has been built and about some of the actors. Could you elaborate on what type of actors participate in the projects and how you attract them to take part?*
In the early days of the FI program, we started with larger players from ICT and agri-food. We also included a large supermarket chain from Spain, for example. That's where we started. Later on, we created an ecosystem for smaller players. That was really highlighted in the third phase of this FI program, in which we established accelerator programs. We managed to have 3 out of 16 of these accelerator programs on agri-food. The other ones were on health, smart cities, and other domains. Three of them were about agri-food as well as logistics. These programs were completely focused on start-ups. Here, we invited all kinds of start-ups to join the ecosystem and to come up with new solutions.

Maria: *When talking about those SMEs and start-ups, what do you think have been the major drivers for these companies to join the projects or ecosystem?*
I think the main driver for these SMEs is that they have limited access to and resources of knowledge to deal with these new technological developments. In these big European projects, they can have access to all kinds of knowledge

and expertise. Both on the technical side, which is usually the most interesting for them, and on the side of business support, as well as governance they can get support for. Usually you see that the latter is not their primary interest, but as soon as they start working on these technological developments, then they realize and feel the need to learn about how to improve their business.

Teun: So, are you suggesting they may not have thought about those kinds of things initially?
Most of these companies are technology-oriented. They have a new idea or new technology and they think they have something unique. Then, through the interactions they have in the network, they find out that their product was not as unique as they initially thought. They suddenly understand that if they want to be unique, they have to look more at their business model and think about how to introduce their renewed products in the market. So, it is not only a matter of a new technology to use, it is a matter of considering how they can find an attractive business model for that technology.

Maria: That makes sense. I would like to move a bit toward governance and coordination. Could you tell me a bit more about how the business ecosystems have been organized and coordinated?
That's an interesting question. I remember when we entered this FI program, we were told by the EC that the project should be industry-led. At that time, we mostly interacted with a large ICT company from Spain, so we told them that we thought they should lead the project. For some reason, they were refraining and didn't want to do it. In the end, we agreed to lead the project ourselves. I think we did a good job there. Overall, I think it worked out well that an institute like WUR took this role. I have had some experience with these kinds of projects in the Netherlands in the past. What you see is that as soon as there is one dominant player, either from ICT or agri-food, coordinating a project, others don't want to join or become suspicious. In the end, I think it was a good decision that we, as an independent institute, were put into this position, and I think we fulfilled this job quite well. We were quite experienced in running big projects and we were well equipped to do that. I think that was appreciated by the partners in the projects, but also by the EC. We built up a good reputation. I think the main thing behind the success of these projects was that we were independent and were not developing products or services in this digital arena. Of course, we sometimes help in developing prototypes and supply knowledge to do so, but in the end, we do not sell those products.

Teun: I can remember from the project proposals that there were all kinds of structures and regulatory bodies involved in the projects, like an executive board or steering committee. Could you elaborate more on that?
Of course. I already emphasized that we had a leading role, but we do need some support. Most of the time, we were the overall coordinator of the projects. As these projects were quite big, for example IOF2020 or SmartAgriHubs,

we specifically needed to arrange a multi-headed coordination team. We were organized into work packages to distribute clusters of similar tasks to other partners. Each work package had a partner responsible for the coordination of the internal work to accomplish. This role was covered by other institutes. I think we carefully selected the partners. In the beginning, we mistakenly met people without knowing whether they were competent at coordinating and managing such complex projects. For some work packages, we immediately found the right collaborator, but we have also concluded that some partners were not the right ones afterwards. I think through past projects we were able to select the most competent people to coordinate these work packages. In addition to these work packages, we always have some advisory or supporting boards, in which you select people from different industries, in this case from agri-food and ICT, that have an independent position in the project. Sometimes it was very hard, given the size of the projects now, to find people who are not involved in one way or the other. We always try to find those people who are somehow unrelated to the project. At key moments in the project, we use to ask for their advice. For example, in IOF2020, we had an important moment in which we had to define and launch an open call. The question was, what should be the themes or priorities of this open call? We came up with a proposal and the advisory board supported and helped us and gave their approval. They came up with some ideas and comments that we implemented. So, support in the management and coordination activities was quite widespread throughout the entire ecosystem. That is basically how we organize such projects.

Teun: So basically, the advisory board is used to create a more objective view of what is going on?
Yes, to create an objective view, and also to get support from those in the other institutions, let's say. This also allows us to show that we are not just doing our own things but are also advised by this external board. Additionally, funding comes from the EC and they give other reviews independently, usually two or three times during the project's duration. We do not have any influence over that. The EC appoints people that are not part of the project to review our results. We also usually spend a day sitting together and critically reviewing the project. Of course, they also have some kind of advisory role. If you do a good job, it is more advice that you get. So, it has a similar function, but we do not choose these people ourselves.

Maria: In your view, what are the principles governing interaction in business ecosystems? Is the project mainly involved with contractual types of governance, like rules, procedures, et cetera, or is it also based on, for example, reciprocity and social norms?
Of course, I think it is a combination of both. You need to have more formal governance mechanisms, but there should also be room for more informal interactions. According to my experience, this often happens. As I said in the beginning, the use cases form the heart of the project. They are formed

by groups of people working on a product or service in a specific domain – dairy, vegetables, meat, et cetera. These groups of people are being monitored and assessed. We have one work package in the project that is devoted to the monitoring and evaluation of these use cases. On a regular basis, people working within these use cases are asked to fill in reports and templates on how they progress; that is the formal part. On top of that, we usually have a discussion about the progress using a teleconference system, since the partners are based in different locations. Additionally, there are several physical or virtual meetings in which we discuss in a more informal way the progress and bottlenecks. As I said, the use cases are also supported on issues related to technical, organizational, and ecosystem-related aspects. Sometimes the interaction we have is more formal. Sometimes it is also very spontaneous and emerges through discussion. So, for example, from a technical perspective we ask the use cases, "What do you see as a reusable component that could be utilized by other use cases?" because this is what we want to facilitate, the diffusion and utilization of a higher level of knowledge and experience that could be beneficial for the other use cases to know. Again, this usually is an exercise we do by using a template that the use cases have to fill in, answering questions about some reusable components. They also get it back, of course. Members of work packages compile everything and we get information back on a regular basis. For example, when two use cases utilize soil sensors to measure soil humidity, we share their experiences and facilitate conversation with other interested partners. This sometimes triggers a process where people within use cases decide to join others for discussion and join efforts to compile data sheets related to soil sensors or a common piece of software. That is happening at all levels. We sometimes have formal events like, for example, annual meetings, in which all work packages and use cases can meet and discuss. We have good experience with the way we have organized the flow of information within these events, in which each work package stays in one place and the use cases come around and discuss particular topics. Sometimes we put them together to discuss different way of working. We try to facilitate interaction and create synergies as much as possible, of course. That is the whole idea of our roles in these projects.

Teun: What are the main tensions that you see arising from the ecosystem or projects? Tensions between actors, for example?
What do you mean exactly with tensions? It varies between the different use cases. Maybe as I already indicated, at the technical level, each use case has its own challenges. That varies among different domains of application – dairy, meat, vegetables, et cetera. You can imagine that the technical challenges for horticulture and arable crop farming are quite different. Of course, at a certain level, there are some similarities, but if you look more at the business modelling aspects, we can see some common problems across the different domains. All these ICT suppliers, in particular the SMEs, are facing very common and similar problems when penetrating the agri-food sector. We could maybe identify that as one of the main

challenges that they are facing. But this is also found among the use cases. If I look at crop farming, there are a lot of use cases working with all kinds of machines and data exchange is taking place between different systems. They are very often sitting together to see if they can create common standard messages or ontologies related to that sector. For example, the dairy trial in IOF2020 decided to sit together to work on their business models because they somehow became aware that they were basically doing the same thing. Indeed, it was very difficult to define a unique business model for each of them. Therefore, they decided to team up a bit and discuss how they could distinguish themselves in the European or global market.

Maria: *I would like to move on a bit further toward technology and change. You already mentioned the different trials, the use cases in meat, vegetables, and dairy, for example. Is there a common denominator for the technologies that have been used within those trials?*

It depends on the level you are looking at. We usually have work packages on technical integration, standardization, and these kinds of things, in which they gradually developed some kind of reference architecture that can be applied to all domains. For this reference architecture, I have a picture in mind. At the bottom, you have the hardware, the sensor devices that measure things, and at the top, you have your applications. In between, you have all kinds of intermediate levels of connectivity and communication that all require standards. There is a lot of overlap there. If you talk about connectivity, they all use the same protocols, like WI-FI and 4G. In the IoT, we are now talking about LoRa. There, we see a lot of potential synergies between the different applications and use cases. If they encounter certain challenges or problems, it is very useful if information and knowledge are shared.

Teun: *Otherwise you cannot finalize the products and you cannot measure the impact.*

If you do not close the loop with the use cases, then you are just delivering a part of the solution, which in practice nobody can use. This is still happening though. For example, farmers are confronted with all kinds of nice products, like sensing devices or drones that can provide a good overview of the field. But if you cannot use it for improving decision-making, why should you buy something like a drone? Of course, there are some farmers that believe in it and are willing to try, but this is rare. You want to have some kind of product or service that you can use, and therefore you have to close this loop. This is always the big question especially in the final part, concerning control. An example I have from precision agriculture is that you can measure and monitor everything on a square meter or even centimeter if you like. I think it is not a big challenge to analyze on that level, but it is questionable whether you can get a machine that is working on this square meter or centimeter level. That is usually a big bottleneck. We still have large machines and they are getting even larger, so that is contradictory.

Maria: When talking about those different technologies that are used in the use cases, to what extent do you think that they have the same of different level of pervasiveness or disruption on the farm's operations?
I think, and this is what I often hear, that technology as such is not disruptive per se; it is more about the business level. People refer to the examples of Airbnb, Uber, and others. If you look at their technology, it was already there for ages. These companies could turn it into a disruptive business model, and I think that also holds true for these projects. As I was saying, putting some new sensing devices in the fields might look nice, or flying a drone over your field, but that is not disruptive in and of itself. If you are really able to have a disruptive business model on top of that, meaning it is a very promising business case for the farmer, as well as for service companies, then you really have disruptive influence. To be honest, I don't see many examples yet in agriculture.

Teun: In the agriculture sector, do you think there is a change in the business model after the adoption of new digital technologies? What are the differences across the different domains you have observed (e.g., dairy, meat, vegetables, etc)?
I think in the end there is not a lot of difference across project domains, but what I could see more in other industries is the development of servitization. That is what I think should happen and is happening in agri-food. I was recently in India looking into these developments. There, I could see a higher concentration of smallholder farms. For them, it is not realistic in the short term to heavily invest in new digital technologies because of cultural and organizational issues. What they have already applied there is that farmers can rent equipment like tractors or smart devices, in a way that it is delivered as a service. I could meet one large company selling the usage of technology as a service or through subscription to farmers.

Teun: So, you think the business model for the farmer remains the same?
Not necessarily. Sometimes farmers, including in India, were able to skip traders, giving them more direct access to the market; for farmers, this is a kind of disruption. Therefore, they were obliged to develop a new business model. Starting with the IOF2020 project, we tried to encourage business actors within the use cases to develop a shared business model. Technology providers, for instance, could incentivize the adoption of new technologies by providing farmers with a well-developed business case. To run a new business with the adoption of new technologies, there should be a cluster of service providers, hardware providers, and infrastructure providers to share the benefits with, as they might be responsible for the related potential improvement. Therefore, they could claim their piece of the cake. For example, if technology providers sell farmers the service of monitoring crops, allowing them to achieve a relevant improvement in the quality of crops themselves and a reduction of other costs, then the revenues from the sales of crops should be somehow shared among all the actors involved, as each of them could have contributed to the delivery of better products. This

is exactly the challenges we have within the use cases, to find a way to achieve a consensus on suitable shared business models.

Maria: You were already discussing integrating business models between different actors. How do you think technology can shape or change the way those actors interact with each other?
I think in that sense you could say that technology is sometimes disruptive because the whole digitalization process provides opportunities for all kinds of players to do something different with more insight on the collected data. Before the usage of digital data and having the possibility to share data with partners, there could be no real business leveraging the useful insights from the company's recent history. But now farmers are also becoming aware that data collected from them can be useful and not only to them. So now they are giving their permission for the usage of their data. They can even give (or sell) their data to another company. This adds an additional source of revenue for SMEs and opportunities for start-ups to do business with the data collected. This is a natural consequence of this current technological development. However, I believe we are still in the early stages of this big transformation. Last Saturday, I participated in a jury for a Hackathon. I observed the case of a company that was able to modify the genotype distribution among their chickens and related hens by acting on some parameters present in the environment, such as humidity and temperature. Relationships between the genotype produced and the changes induced in the environment were initially not stored and shared but given the availability of new technology they decided to start storing their data and to give their analysis to a start-up. This is an example of how the nature of business is changing.

Teun: That is actually quite an eye opener. Do you have any ideas on or have you observed any patterns of change about how business ecosystems or projects have evolved over time? You were already elaborating about how it expanded over the years. Could you go more into detail?
I can easily think of my recent battle for the development and availability of relevant digital data in the agri-food sector. Big agricultural companies were the first to become aware of the relevance of having available huge amounts of data. Soon after, large retailers and food processors realized that they could use the acquired data to improve their products. Only afterwards could they think that this amount of data could become another source of revenue for them. Other companies started recognizing the potential of this data and tried to use it to improve their performances. As natural consequence of this, the problem that emerged was who had to own this data and who can access it. The issue has been solved by imitating what occurred in other sectors. For instance, in the case of Facebook and Google, we as citizens and users now have more control over our data, or at least we have more opportunities to exercise control. Relevant examples of this is when we give companies our bank account for accomplishing various internet transactions and then we are asked if these

companies can develop some services related to our personal data. Similar dynamics are also occurring in agri-food sector.

Another relevant element affecting the dynamics of business ecosystems is the presence of many additional players. These new business actors tend to be smart and visionary. Among them, there were many start-ups that were able to identify opportunities. Since they came onto the scene, these start-ups have been supported in their development through accelerators and incubators established with the help of European funding. Other start-ups were instead funded and supported by large companies. Some start-ups suddenly became big companies even though they did not come from large, traditional enterprises. In addition, we can also see venture capital participating in the development of our focal ecosystems, resulting in good examples of start-ups quickly becoming big players.

Due to digitalization, many different players operate within the same arena. Some of them can compete with more traditional players. The establishment of a common digital infrastructure could constitute a sort of playing field, allowing them to play with standards, open infrastructures, and innovate. These dynamics are gradually evolving. Of course, big players are trying to maintain their positions. However, in the agri-food sector, even large companies can have access to only one piece of data. For example, there are some companies that have access to machine data, but they do not always have access to crop or soil data or market data.

Maria: Is there is a need for one actor in the business ecosystem to adapt to the other actors?
Sure. As I was just saying, there were traditional ICT providers, for example, of farm management systems. Before the digital transformation, a few local players distributed farm management systems within a geographical area. With the digital transformation, many other players entered the field, even from other sectors. Some of them, more technologically equipped, asked farmers to provide them with their data so that they could deliver more sophisticated services. To avoid a lock-in effect, farmers have started asking for the implementation of an open interface in the system. Perhaps this is fair since, in the end, these companies are using their personal data; farmers therefore have the right to shape the development of the system. This is an example of how traditional farm management systems have to adapt to this new technology evolution in which farmers are playing a leading role.

Maria: Did you find some problems in the communication between ICT players, who were using specific language, and farmers? If so, how did you tackle this problem?
It is a very common problem. The ICT and agri-food sectors are quite different, requiring different skills and knowledge backgrounds. In coordinating our business ecosystems, we have tried to cope with the difficulty that partners are facing in understanding one another. We have attempted to translate relevant messages and we have always asked for simplicity in the report and in any other

form of communication toward others. So, the difficulty of having a common language is a really central problem and triggers our actions in the organization of our coordination activities. In the role of facilitator, we are trying to translate messages and provide a glossary of terms and abbreviations that are used by the parties involved. The issue is that we have to coordinate different actors within each use case. At beginning of their work together, they had a difficult time communicating with one another. Nevertheless, they were called to work together heavily on designing, implementing, and evaluating the product or service that technological providers were developing. But it was clear that they could not understand each other, so we had to work a lot on this aspect.

We only have a few farmers involved in each use case, while the rest of the community of farmers does not participate. For them, we still need to facilitate the translation of meaning. We are trying to cope with this problem by emphasizing the relevance of having a clear user interface, a good user experience for any type of technology involved. Related to this, I think the most important aspect to tackle is that farmers understand the usefulness of technology. If the technology is also difficult to use, they are not going to adopt it in their daily operations and any possibility to have a new business case out of the application of such technology will disappear. For instance, if you have an app installed on your smartphone and you have a good experience with it from the very beginning, then you will use it again and again, but if your first experience with it is bad, then you will most likely delete it.

Teun: Let's move a bit toward the transition and evolution of ecosystems. We were already discussing the role of the EC in the direction of the projects that you are working on. To what extent do you think, or have you experienced that other elements of the institutional environment have influenced the transition or evolution of these projects?
I think it is interesting to say that the EC consists of different directorates general (DG) as you call them, like different ministries in a country. Most of these projects were supported by DG Connect, which is the ICT ministry, let's say. They were interested in creating this generic system between all the different domains. At first, agri-food was not on the radar at all, so we had to fight to have it on the map. Then you also have DG Agri, which is usually involved in the common agricultural policy. They gradually became more interested in the digitalization of agriculture and now they are trying to combine it with this common agricultural policy. This happened with the IOF2020 project, which is why it has become so large. At a certain point, DG Connect decided they wanted to have a large IoT program for different sectors. At that time, in the agri-food sector there was a perceived need to start digitalizing, so they agreed to have a large-scale IoT pilot in agri-food as well. DG Connect provided 15 million euros. Later, DG Agri wanted to join and provided an additional 15 million euros, making IOF2020 the biggest EU project ever. It is even bigger than the automotive or smart cities EU projects, actually. In addition to the EC, there are all kinds of lobby organizations from the farmers' side, like CEMA, which is the association for agricultural machinery manufacturers, as well as big

companies working on digitalization on their own. Of course, they are trying to influence the agenda of DG Connect and DG Agri. There is a whole playing field with different large lobbying organizations trying to influence the agenda. This is also true for us as a research and development institute. We want to have successful proposals for these kinds of projects, but we somehow need to get some support. This means that we have to consider what these large companies and associations are saying because they are often involved in the evaluation of EU proposals. That is how it works, I think.

Maria: *To what extent do you think things are missing for the ecosystem to sustain itself?*
Good question. One of our early projects in these FI projects was FIspace, which developed a business-to-business (B2B) collaboration platform, working like a social network, like LinkedIn or Facebook. After this project, we tried to transition into the FIspace Foundation to be able to continue after the project. At that time, we did not succeed. What worked was that FIspace was based on FIware, a framework of open source platform components used to accelerate the development of Smart Solutions, which was developed for the other sectors. That took off. They could transform the effort of FIspace into the FIware Foundation, and it is still running quite well I think. Agri-food still is one of their main themes, so although we did not succeed in having this FIspace Foundation, somehow we were able to continue part of the work that was done.

With the latest program, SmartAgriHubs, which is a very close follow-up to IOF2020, we really want to target the fragmentation of these developments in Europe. The challenge is that, on the one hand, to overcome this fragmentation we want to avoid the same products or services being developed in different locations in Europe, reinventing the wheel. On the other hand, we do not want to put everything into one big organization. We want to maintain the local or regional level because in the end, every regional context is different. We have to implement things in a different way. At the same time, we want to see how others can connect with each other. That is the big challenge for SmartAgriHubs. The title of the project is Connecting the Dots for Digital Transformation in Agri-food. Here we want to use local or regional energy but utilize the power of the network that we created. For example, if there is a development in the precision fertilization of potatoes in Romania, players can look into the network to see if there are any other digital innovation hubs, competence centers, or experiments going on in the same area. Can we learn from them? Can we perhaps use some components or technical parts? Can we learn from the business model they developed? It's about how to connect that with each other using the network. The big challenge is that we set this up in the project, and the project will end in 2022. Then what? Will everything fall apart, or do we need to have another project again? We consider whether we can maintain this SmartAgriHubs network in another way, perhaps again in some kind of foundation, hopefully successful. Maybe we will find another construction or governance mechanism, but we still have to find out. We started with this project not too long ago, so we are beginning to think about these issues. However, it is not our main concern now.

Teun: *What is your experience with former projects, when they ended? These projects were based on temporary consortia. What happened to these ecosystems that were established during these projects?*
I think until now we have been successful in setting up several EU projects for driving the application of digitalization in the agri-food sector and in coordinating them. We could indeed rely on a core set of partners that grew in numbers over the years. Through our contacts and structure, we could keep the networks cohesive. But this is a set-up which is not sustainable in the long run once projects are over. However, we believe that the presence of associations, standardization organisms, and umbrella organizations that are currently involved will act to take the lead once we finish. These organizations have lengthy experiences and are supposed to survive, so they might have an interest in letting the ecosystems survive too. Additionally, part of the knowledge developed, and their newly produced technological components have already been integrated into these organizations. In that sense, we were already successful in sustaining the results of the projects, and also partly of the business ecosystems established. As I said, we should also aim for a SmartAgriHubs Foundation afterwards.

Maria: *What is your vision on the future of digitalization in agri-food?*
As I said, digitalization is still in an early phase, so I think much more digital data will become available, leading to multiple uses. Not only to support, for example, farm management or logistics management, or even consumer decision support, but for many other purposes as well. For example, monitoring public objectives like climate change or sustainability in general. Additionally, I think we will increasingly use this data for research and development. There is a mutual dependency between the development and applicability of technologies like big data, IoT, or blockchain. As a research organization, we can better develop these technologies when we have more access to this kind of digital data. I think we can expect this much more in the future. I think we will see this in advanced applications or even artificial intelligence. Everybody is talking about it now and it is still in its infancy. Besides, I expect there to be more autonomy in farm management, like autonomous vehicles. In combination with artificial intelligence, more decision will be made by computers. Again, I think this is widely applicable to other sectors and domains, but I expect the same things will happen in agri-food. The way our food is produced will change. Decision making will shift to a higher level, a higher intelligence level. Farmers are now busy with operational management. In the distant future, I expect that farmers can focus more on market or business decisions, for example.

Teun: *Allowing farmers to work on a more abstract level.*
Exactly.

Maria: *All right. I would like to thank you very much for this interview.*

6 Research insights on the governance and dynamics of IOT business ecosystems

Teun Gilissen, Maria Carmela Annosi, and Wilfred Dolfsma

Introduction

Innovation management and the spread of new technology are one of the most frequently studied areas in management literature (Dodgson, Gann, and Phillips, 2013; Rogers, 1995). This is not surprising, as it is often argued that innovation is an essential means for organizations to create a sustainable competitive advantage and survive in the long run. Several meta-analyses showed that innovation is positively related to organizational performance (e.g., Bowen, Rostami, and Steel, 2010; Rosenbusch, Brinckmann, Bausch, and Bausch, 2011). Studies on a macro-level indicate that innovation drives long-term economic growth (Kraemer-Mbula and Wamae, 2010). In addition, it is argued that innovation contributes to environmental sustainability, employment, and even social well-being (Dodgson et al., 2013). Hence, it is evident that innovation is essential for organizations to achieve a sustainable competitive advantage (Barney, 1991).

Nevertheless, innovative organizations cannot evolve in a vacuum, especially while considering small and medium-sized enterprises (SMEs). In European economies, SMEs account for 99.8% of all business enterprises and 66.4% of all employment. Nearly all these SMEs are micro SMEs, which consists of less than ten employees, a turnover of less than €2 million a year, and a total balance sheet of less than €2 million (European Commission, 2018).

As suggested by Zahra, Neubaum, and Naldi (2007), SMEs are dependent on various types of resources, and in particular on their ability to acquire and transfer knowledge. SMEs often encounter limitations in accessing these different types of resources. As a consequence, this may influence the innovative performance of SMEs (Van Burg, Podoynitsyna, Beck, and Lommelen, 2012). A means to overcome these limitations is to attract resources externally, relying on a network of partners, suppliers, and customers (Moore, 1993). Scholars have often pointed out the critical importance of such interorganizational relationships to achieve competitive advantage (Cao and Lumineau, 2015). These partnerships often involve cross-industry collaborations (Turber, Vom Brocke, Gassmann, and Fleisch, 2014). As a consequence, organizations become connected with each other through both technical and business ties, thereby increasing interdependency (Westerlund, Leminen, and Rajahonka, 2014). It is argued that the structure of such social networks in the agri-food sector impacts the spread of innovations (Deroïan, 2002).

This perspective of organizations and their relationships is in contrast with the classical view of organizations. Traditionally, organizations were seen as single actors within an industry, battling for market share with their direct competitors (Moore, 1993). In this view, actors are organized as value chains in which the value creation process is linear and flows upstream to downstream (Clarysse et al., 2014). However, this view seems outdated and limited, especially when considering the dynamics of the business environment. To cope with this issue, Moore (1993) suggested an alternative view, in which organizations are considered part of a business ecosystem that crosses industries. Business ecosystems can be conceptualized by comparing them to their biological counterpart. Parallel to species in biological ecosystems, organizations interact with each other in a complex environment, thereby constituting networks of interdependencies to function (Iansiti and Levien, 2004b). According to Moore (1998, p. 168), business ecosystems consist of:

> Communities of customers, suppliers, lead producers, and other stakeholders – interacting with one another to produce goods and services. We should also include in the business ecosystem those who provide financing, as well as relevant trade associations, standards bodies, labour unions, governmental and quasigovernmental institutions, and other interested parties.

Moore (1993) suggests that every business ecosystem goes through four distinct phases. These include foundation, expansion, leadership, and self-renewal or death. Each stage is characterized by different strategies and types of collaboration that organizations within an ecosystem can employ. Although these phases have been addressed conceptually, not many studies have empirically examined the dynamics of business ecosystems. Based on a review of empirical literature, Provan, Fish, and Sydow (2007) raised some gaps in our understanding of business ecosystems (or whole networks, as they call it), which still seem relevant. One of the research gaps they identified relates to the dynamics and development of business ecosystems, as there is no clear empirical evidence on how these ecosystems evolve over time (Provan et al., 2007; Zaheer et al., 2010). Scholars have often called for a more dynamic view of network research to cope with the criticisms on the static nature of these studies (Ahuja et al., 2012; Moretti, 2017). However, this call has not yet been addressed sufficiently.

A related but different issue that is raised is whether, how, and when the governance of business ecosystems change over time (Provan et al., 2007; Provan and Kenis, 2008; Reuver and Bouwman, 2012). There is a lack of clarity with regard to the interplay between different types of governance mechanisms – contractual versus relational governance (Williamson, 1991; Zaheer and Venkatraman, 1995) – as well as when different mechanisms are used. A longitudinal perspective is required to capture all relevant characteristics and dynamics of governance mechanisms in business ecosystem (Roehrich, Selviaridis, Kalra, Van Der Valk, and Fang, 2019).

This chapter further deepens our understanding of business ecosystems in general, the evolution of business ecosystems, and business ecosystem governance. Based on this literature review, several knowledge gaps are identified, and future research areas are suggested.

Business ecosystems

Business ecosystems can be compared to its biological counterpart, in which all actors are interdependent of each other.

Previous research has focused on three characteristics of business ecosystems (Li, 2009; Scaringella and Radziwon, 2018). First, a business ecosystem consists of a loose network of actors with various backgrounds, which are affected by the creation and delivery of an organization's offerings. As mentioned above, this can vary from suppliers, distributors, outsourcing firms, producers of related products, or other stakeholders. The second characteristic is the use of a platform, which refers to products, services, or technologies that other actors in the ecosystem can build upon (Li, 2009; Scaringella and Radziwon, 2018). The third characteristics refer to the scope and evolution of the business ecosystem. Organizations within a business ecosystem coevolve through collective innovation and value creation, in a way that the synergetic value is greater than the sum of the parts (Li, 2009). Furthermore, scholars pointed out three dimensions of the business ecosystem: context, configuration, and cooperation (Li, 2009; Rong, Hu, Lin, Shi, and Guo, 2015).

Iansiti and Levien (2004b) argue that a large number of loosely interconnected actors, dependent on one another for their mutual performance, are essential for business ecosystems. Each actor provides particular contributions for collective value creation and influences the business ecosystem differently. Three types of ecosystem players can be distinguished, based on the dynamics of an industry and the complexity of relationships (Iansiti and Levien, 2004a):

Niche player: an organization that develops its own specialization to differentiate from competitors. This strategy is appropriate when an organization faces rapid and constant change, and when the relationships with other ecosystem actors are relatively simple.

Physical dominator: an organization that operates in a mature industry and a relatively stable environment. A complex network of external assets and relationships is in place. Physical dominators may choose to acquire their partners in an attempt to control essential assets.

Value dominator or keystone player: an organization that is the center of a complex network of assets-sharing relationships within a turbulent environment. Keystone organizations that carefully share (the wealth generated by) their assets can stimulate ecosystem innovation to deal with disruption in the environment.

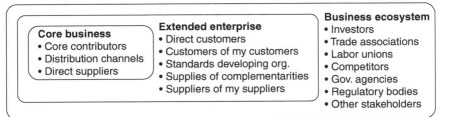

Figure 6.1 Layers of a business ecosystem. Based on Moore (1996).

Ikävalko, Turkama, and Smedlund (2018) suggest that value co-creation is the interaction between organizations and their customers, suppliers, business partners, competitors, and other third parties. They suggest alternative archetype roles within an ecosystem, which include ideators, designers, and intermediaries. An ideator brings knowledge to the ecosystem in a one-way knowledge flow, providing input for service innovation. A designer integrates existing knowledge components into the ecosystem to develop service innovation, using reciprocal knowledge flows. An intermediary acts as a broker and communicates knowledge to multiple ecosystems. The roles differ in operating logic and ecosystem activities (Ikävalko et al., 2018). Figure 6.1 illustrates the different layers of a business ecosystem along with the concerned players (Moore, 1996). The boundaries of business ecosystems extend the traditional view of the *core business*, including distribution channels, suppliers, and an organization's core capabilities, and the *extended enterprise*.

Indeed, Moore (1996) suggests that it is essential for organizations to consider their surrounding environment, including all factors and firms that may influence this environment. According to Moore (1996), an even more holistic business ecosystem should include investors, trade associations, labor unions, competitors, government agencies, regulatory bodies, and other stakeholders that can affect or are affected by other actors within the business ecosystem.

To further clarify Moore's model, we now exemplify how these general blocks of business ecosystems are translated into the agri-food context. The core business of a farmer consists of, for example, fertilizer and fodder producers, machinery producers, and other input providers, which allows employees to produce goods and add (contribute) value. The core business also includes distributors of input (e.g., seeds, fertilizer, farm equipment) and output (e.g., dairy, raw vegetables, or animals). In the extended enterprise, a broader view of all stakeholders involved is considered. The direct customers of a farmer are dependent on the output and the complexity of the value chain. Some examples include processors, traders, and small businesses, like restaurants, markets, or cooperatives. Again, the type of customers is dependent on the output and value chain, but could include secondary processors, traders, markets, logistic service providers, or retailers. Standard-developing organizations refer to organizations that develop industry or production practices, for example on sustainable production. As complementary products or services may create synergy effects, they should be taken into account. For instance, seeds and fertilizer are complementary to each other because planting seeds requires fertilizer, creating interdependency. Lastly, the suppliers of a farmer's own suppliers could refer to raw material providers or producers or other parties providing inputs to a farmer's suppliers.

Within the agri-food context, both general and specific agri-food investors can be identified, including banks, angel investors, venture capitalists, or family. Numerous trade associations are involved in the agri-food sector, including the Agri Food Chain Coalition (AFCC), the European Liaison Committee for the Agricultural and Agri-Food Trade (CELCAA), and the International Agri-Food Network (IAFN). In addition, national and international labor unions, like the European Federation of Trade Unions, in the food, agriculture, and tourism sectors, may affect the agri-food business ecosystem by supporting sustainable development of agri-food policy. Competitive farmers definitely affect a farm, for example in terms of supply. However, organizations may collaborate with their competitors as well, creating a coopetition structure (see cooperation or coopetition in the following section). Government agencies and other regulatory bodies may impose legal boundaries, thereby impacting agri-food businesses. Other stakeholders that affect or may be affected by an organization should be taken into account as well.

Emerging collective actions within business ecosystems: cooperation or coopetition

Cooperation links the context and configuration and illustrates the roles that stakeholders perform. Within a business ecosystem, both collaborative and competitive relationships can be established, creating a coopetition structure (Clarysse et al., 2014; Moore, 1993). Organizations can exploit their interdependencies and create competitive advantages through collaboration in a business ecosystem. They co-evolve their capabilities and roles and align themselves with the direction set by one or several central actors (Clarysse et al., 2014). The emphasis of this perspective of business ecosystems, which Adner (2017) calls *ecosystem-as-affiliation*, is on the breakdown of traditional industry boundaries, interactions between ecosystem members, and the increasing interdependency among these players. Actors or organizations are of central concern. Complementary, Adner (2017) describes the *ecosystem-as-structure* perspective, which starts with a value proposition and aims to find actors that could contribute or add to the value proposition. This perspective focuses more on activities, which is illustrated in Figure 6.2. The output of upstream suppliers contributes to or adds value and serves as input for the focal actor. Other organizations within the ecosystem can offer complementary products or services in order to maximize value (Adner and Kapoor, 2010; Moore, 1998). Drawing on the latter perspective, players typically cooperate, compete, and co-evolve capabilities around a core innovation

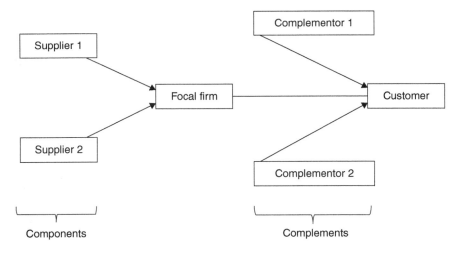

Figure 6.2 Generic schema of an ecosystem. Adapted from Adner and Kapoor (2010).

(Moore, 1993). These innovations are often managed through an open platform, which is initiated by the keystone organization. An open or external platform can be defined as "products, services, or technologies developed by one or more firms, and which serve as foundations upon which a larger number of firms can build further complementary innovations and potentially generate network effects" (Gawer and Cusumano, 2014, p. 420).

These platforms can vary from a physical asset, like specific manufacturing capabilities, to intellectual assets, like Windows' software platform (Iansiti and Levien, 2004a). In such a self-contained system, actors create, generate, and produce new output, structure, or behaviors without any input from the focal actor. Overreliance on such a system may lead to uncontrolled creative output, which may damage the health of the ecosystem (Wareham, Fox, and Cano Giner, 2014). This emphasizes the urgency of adequate business ecosystem governance, which is further elaborated in the following section.

Governance mechanisms within business ecosystems

The complexity of interactions that arise within a business ecosystem requires governance mechanisms to manage the ecosystems effectively. Generally speaking, governance determines who has power, who makes decisions, how other stakeholders make their voice heard, and how account is rendered (Institute on Governance, 2019). Transaction cost economics (TCE) has commonly been used to understand how organizations construct governance arrangements with organizations they cooperate with. Organizations typically focus on minimizing cost, while ensuring delivery of the desired quality, price, and quantity, of the service of a supplier. According to TCE, the governance features of interorganizational relationships are matched with known exchange hazards, such as asset investments, performance measurement, or uncertainty (Poppo and Zenger, 2002; Williamson, 1991). Formal contracts are constructed that represent promises or obligations to perform particular actions in the future, as a means to control for opportunistic behavior (Vlaar, Van Den Bosch, and Volberda, 2007; Wang and Wei, 2007). Based on different forms of contract law, Williamson (1991) identified three forms of governance within organizations. Market and hierarchy are two poles of a continuum, with a hybrid form in between. Markets are spontaneous coordination mechanisms driven by the self-interested actions of individuals and firms. This pole is characterized by choice, flexibility, and opportunity. Prices capture all relevant information necessary for exchange and actors bear no dependency in relation to one another. The hierarchy pole is characterized by clear departmental boundaries, authority, reporting mechanisms, and formal decision-making procedures. Work activities are highly interdependent and require social interaction. In between is the hybrid form of governance, in which players maintain autonomy but develop contracts to manage bilateral dependency (Powell, 1990; Williamson, 1991).

Contractual and relational governance

It is argued that Williamson's market-hierarchy continuum is too static and does not fit most economic exchange properly. His view implies that organizations are separate from markets or more broadly, the societal context. Therefore, others suggested a distinct third form of governance instead of the hybrid form in between (e.g., Dolfsma and van der Eijk, 2010; Powell, 1990). In an attempt to overcome the limitations described above, Powell (1990) suggested a network form of governance. This form perceives organizations as interdependent actors which have complementary strengths and strive for open-ended mutual benefits. In contrast with Williamson's (1991) view, relationships between actors within a network are believed to be not only formally maintained, through contracts, rules, and regulations, but to a larger extent informally, through the structure of the network and norms of reciprocity and trust (Larson, 1992; Provan et al., 2007). These different types of governance are also referred to as contractual and relational governance (Roehrich et al., 2019; Zaheer and Venkatraman, 1995), although not all relational governance mechanisms are informal and vice versa (Moretti, 2017). Alongside contracts, rules, and procedures, contractual governance can also include common staff (spanning organizational boundaries), linking-pins, or joint network information systems (Moretti, 2017). Some scholars argue that contractual governance mechanisms consist of a control and coordination aspect, both having different characteristics. The control function refers to protecting against potential opportunism, whereas coordination is about dividing roles and responsibilities, communication, information sharing, and joint problem solving (Roehrich et al., 2019).

Relational governance is a response to the observation that interorganizational exchanges are often repeated exchanges embedded in social relationships rather than in formal contracts. These relationships can be established by repeatedly using basic coordination mechanisms, such as communication, decisions, negotiations, and meetings between actors (Moretti, 2017). Other mechanisms like group norms, reputation, and peer control may arise from repeated interactions as well. These social processes require both trust and relational norms, such as flexibility, solidarity, information exchange, cooperation, and mutual adaptation (Poppo and Zenger, 2002). Trust is defined here as "a psychological state comprising the intention to accept vulnerability based upon positive expectations of the intentions or behavior of another" (Rousseau, Sitkin, Burt, and Camerer, 1998, p. 395). Adherence to relational norms helps reduce transaction costs (Dolfsma and van der Eijk, 2010; Jones, Hesterly, and Borgatti, 1997). A general belief in organizational network research is that economic transactions occur mostly within the context of social relationships, which is known as the embeddedness principle. In other words, embeddedness involves the overlap between social and economic ties both within and between organizations (Kilduff and Brass, 2010).

Early studies argued that contractual governance negatively affects relational governance, as the use of detailed contracts would signal a lack of trust (Roehrich et al., 2019). It is now believed that contractual and relational governance are complementary (Cao and Lumineau, 2015; Poppo and Zenger, 2002) rather than substitutive (e.g., Vlaar, Van Den Bosch, and Volberda, 2007) to one another. Some scholars argue that the control and coordination function of contractual governance interplay differently with relational governance (Roehrich et al., 2019). They suggest that control signals a lack of trust, whereas the coordination function creates common knowledge and therefore aids the development of trust. According to Cao and Lumineau (2015), the relationship between contractual and relational governance is moderated by the institutional environment and the type and length of the interorganizational relationships. However, most of these studies focused mainly on governance mechanisms in dyadic relationships or formal relationships through for instance alliances (Cao and Lumineau, 2015; Roehrich et al., 2019). Hence, it is ambiguous whether this holds for business ecosystem governance as well.

Forms of governance

Provan and Kenis (2008) identified three forms of network governance that are used to manage an interorganizational network. The three forms vary on two dimensions: the level of centralization and whether the network is participant or externally governed. Shared governance is the simplest and most common form in which the network is decentralized and governed by the members themselves. Governance can be either relational – through the structure of the network or norms of reciprocity and trust – or contractual, through contracts, rules, and regulations. Shared governance is dependent on the involvement and commitment of all actors or a significant part thereof. Members of the network are responsible for the operations as well as management of internal and

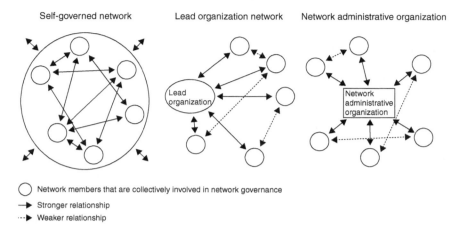

Figure 6.3 Three governance forms of whole networks. Adapted from Raab and Kenis (2009).

external relationships. Network activities and decisions are managed jointly. Shared governance is likely to be effective when trust is pervasive and widely distributed among the network (Provan and Kenis, 2008).

Lead organization governance is a more centralized form of governance that copes with the inefficiencies of shared governance. Here, a lead organization – or keystone organization (Iansiti and Levien, 2004b) – coordinates all network-level activities and key decisions. This type of governance typically occurs in vertical, buyer–suppliers relationships, or multilateral networks in which there is a dominant organization that has sufficient resources (Provan and Kenis, 2008).

Network Administrative Organization (NAO) governance is a type of governance in which a separate administrative entity is set up to govern the network and its activities. Like lead organization governance, an NAO model is centralized, in which the NAO acts as a broker between parties in the network. Contrary to a lead organization, an NAO does not offer products or services other than the coordination and management of the network. Figure 6.3 illustrates the differences in governance between the three forms (Raab and Kenis, 2009).

Business ecosystem dynamics

Business ecosystems evolve over time and therefore it is necessary to use a dynamic point of view when analyzing business ecosystems (Battistella, Colucci, de Toni, and Nonino, 2013). Understanding how networks change and evolve is one of the most relevant challenges for network research at this time (Moretti, 2017). As suggested by Kilduff and Tsai (2003), two ideal types of change processes can be distinguished in network development: serendipity and goal-directedness. These two types differ in structural dynamics and how they operate, thus influencing evolution of the ecosystem over time. In the serendipitous process, it is assumed that network change is based on spontaneous interactions between actors within a network that do not have a shared goal. Network growth is slow and predictable and evolves through dyadic interactions. Structural holes and subgroups emerge as there is a decentralized structure and no leading organizations. Change based on goal-directed processes emphasizes that networks of organizations emerge to fulfil a shared goal. The network evolves through rapid collection of network members around the shared goal or through changes in the common goal. An administrative entity is set up to coordinate the network's activities, which may be either a member of the network or a separate player (see Forms of Governance) (Kilduff and Tsai, 2003; Moretti, 2017). Although there may be differences between serendipitous and goal-directed processes, all business ecosystems evolve within lifecycles (Moore, 1993), in response to changes in its business environment (Rong et al., 2015).

Current research gaps and research implications

As discussed, business ecosystem governance describes what mechanisms are used to govern the overall ecosystem activities. Forms of governance are required to ensure that actors engage in collective and mutually supportive action, address conflicts, acquire and utilize resources effectively and efficiently, and ultimately reach network effectiveness (Provan and Kenis, 2008). It can be seen as the basic engine of network functioning and the foundation of interfirm cooperation (Moretti, 2017). Vangen, Hayes, and Cornforth (2015, p. 1244) provide a useful explanation to further define this concept as "the design and use of a structure and processes that enable actors to direct, coordinate, and allocate resources for the collaboration as a whole and to account for its activities." As the governance mechanisms are likely to differ when considering the business ecosystem as a whole, the following areas of investigation are proposed in this study:

> Investigate to what extent both contractual and relational governance mechanisms are adopted to govern a business ecosystem;
>
> Investigate the different functions of contractual governance in relation to relational governance;

Does the control function of contractual governance mechanisms act as substitutes to relational governance mechanisms within business ecosystem governance?

Does the coordination function of contractual governance mechanisms act as complements to relational governance mechanisms within business ecosystem governance?

Business ecosystem lifecycle

According to Moore (1993), every business ecosystem goes through four distinct stages, including foundation, expansion, leadership, and self-renewal or death. The first stage involves defining customer wants or a value proposition through co-evolvement. It often pays to cooperate, as this may prevent actors to help other emerging ecosystems. Additionally, a leading organization must emerge to initiate rapid, ongoing innovation that drives the entire business ecosystem. Expansion is about capturing territory by competing against other ecosystems, stimulating demand and adequate supply. Two conditions are required for business ecosystems to reach this stage. First, the business concept is valued by a large customer base. Second, the concept can be scaled up to reach this market. Dominant players can exercise their power and resources in marketing and sales but generate economies of scale in production and distribution as well. In this stage, it is essential to stimulate market demand without exceeding the corresponding supply (Moore, 1993).

In the leadership stage, the investment directions and technical standards of the business ecosystem are determined, and a robust community of suppliers is created. Organizations could consider taking over activities from their suppliers or customers by acquiring them, allowing them to maintain control over key value-adding elements and increase bargaining power. This requires that ecosystems have strong enough growth and profitability, as well as the value-adding components and complements to become stable. The final stage, self-renewal or death, occurs when mature business ecosystems are threatened by new innovations and ecosystems. It is essential for the long-term success of a business ecosystem and its ability to renew itself to lead successive generations of innovations.

In all stages of a business ecosystem, co-evolution is an essential feature (Moore, 1993). Including lifecycle stages helps to demonstrate firms' statuses at different stages (Rong et al., 2015). The ecosystem life cycle construct would benefit from a more thorough body of empirical evidence to support Moore's work (1993) and to address the research gaps suggested by Provan et al. (2007). Therefore, the following areas of research are proposed:

Verify whether each business ecosystem goes through four distinct and predictable stages which can clearly be identified as foundation, expansion, leadership, and self-renewal or death.

Investigate to what extent actors within a business ecosystem employ distinct strategies in each stage of the business ecosystem cycle.

Dynamics of governance mechanisms and governance forms

Some studies have addressed some elements of the dynamics of business ecosystems and their governance mechanisms. For instance, in a case study, Ness and Haugland (2005) found that governance mechanisms in interfirm relationships developed in multiple phases, moving from market and hierarchy-based governance toward relational-based (network) governance. De Reuver and Bouwman (2012) studied the relationship between innovation phases and governance mechanisms. They found that hierarchy-based governance was used most frequently in the development and implementation phase. Market-based governance was mostly used in the implementation phase. The use of trust-based governance increased over time during the three phases, being strongest in the commercialization phase. As existing literature provides a blurred picture of the dynamics of governance mechanisms, the following areas of research are proposed:

> Investigate whether the development of governance mechanisms show a clear pattern over time;
> Investigate whether the development of governance mechanisms is related to stages of the business ecosystem cycle.

Provan and Kenis (2008) argue that governance forms most likely shift from shared governance to lead organization governance or NAO but not the other way around. Shared governance is the most flexible and adaptable governance form. This flexibility can be used to respond quickly to threats in the environment as well as to opportunities. However, stability may be required to use a network efficiently and to create long-term relationships between organizations. To enable this, the governance form may be transformed into lead organization or NAO. Moreover, governance may shift from lead organization to NAO relatively easy as well, as contingency factors are likely to change in favor of NAO. However, as these two forms have become more stable and less flexible, it would be difficult to transform back into a shared governance form (Provan and Kenis, 2008). According to Provan and Kenis (2008), a clear pattern can be identified with regard to the governance forms. However, it is not clear whether this is related to the stages of the business ecosystem cycle. Therefore, future research could focus on the following research gaps:

> Investigate whether the development of governance forms of business ecosystems show a clear pattern over time;
> Investigate whether the development of governance forms of business ecosystems is related to stages of the business ecosystem cycle.

External factors

It seems that contextual elements may influence governance mechanisms and forms as well, both through external factors and internal contingencies (Bryson, Crosby, and Stone, 2015). According to contingency theory, the performance of organizations depends on whether they can adapt its characteristics to environmental factors (Donaldson, 2001). These factors mainly focus on the macro environment of organizations (Fernández-Olmos and Ramírez-Alesón, 2017).

Collaboration literature has focused extensively on external factors to explain successful collaboration between organizations, for example through alliances. As collaboration is essential for a business ecosystem to thrive, it seems worth exploring these factors from a broader ecosystem perspective. Thompson (2003) suggests that external factors of collaboration can be political, economic, and socio-cultural.

Amongst others, government policies, regulations, and mandates could impact the governance forms that are used in cross-sector collaborations. For instance, mandates can provide one actor with more authority and control compared to other players in a business ecosystem (Bryson et al., 2015). Economic factors may influence the success of interorganizational collaborations as well. For example, Fernández-Olmos and Ramírez-Alesón (2017) found that the relationship between technological collaboration and innovation performance is moderated by stages in the macro-economic cycle. Changes in the business climate as well as instability of the climate may require different collaboration strategies. Besides, it may affect the business opportunities available (Patel, Pettitt, and Wilson, 2012). Socio-cultural factors reflect the national or regional level, and not the organizational or team culture that is studied quite often (Patel et al., 2012). National culture influences the formation and level of trust in society. Besides, it impacts the commitment in interorganizational relationships and knowledge sharing and problem solving in collaborations (Griffith, Myers, and Harvey, 2006). A culture that is characterized by individualism may influence the extent to which actors within an ecosystem want to engage in collaborative behavior with others, compared to a country that has a collectivistic culture (Thompson, 2003). Moreover, Delerue and Simon (2009) argue that cultural factors significantly impact risk perception of interorganizational relationships, which may affect the extent of contractual as well as relational governance mechanisms applied. To summarize, the following areas of research are proposed:

> Investigate to what extent the development of governance mechanisms and governance forms is influenced by external, macro environmental factors;
> To what extent is the development of governance mechanisms and governance forms influenced:
>
> - by political factors?
> - by economic factors?
> - by socio-cultural factors?

Internal contingencies

Provan and Kenis (2008) used an internal perspective to predict the adoption of a specific governance form over the others. Changes in internal contingencies may cause an ecosystem to adopt a different governance form to ensure alignment and thus the effectiveness of an ecosystem. Provan and Kenis (2008) argue that the adoption of a governance form depends on four structural and relational contingencies: network size, the nature of the task, goal consensus, and (the distribution or degree of) trust among members. These contingencies may also relate to some important features within a business ecosystem, like information sharing and collaboration, and therefore impact governance mechanisms employed.

As the number of players within a network increases, the number of potential interactions increases exponentially (Provan and Kenis, 2008). This issue arises in every business ecosystem that is growing, as coordinating the needs and activities of actors is characterized by increased complexity. When the network size reaches a certain level, centralizing the ecosystem governance will increase effectiveness. It is thus more likely that large business ecosystems are governed through a lead organization or NAO. Collective learning is essential in network governance and consists of information sharing, deliberation, and resilience. Newig, Günther, and Pahl-Wostl (2010) argue that network size is positively related to information sharing and the resilience of a network, whereas they expect a convex curve relationship between size and knowledge deliberation (i.e., genuine exchange of ideas and topics).

Organizations become part of a business ecosystem for various reasons, but ultimately they do in an attempt to achieve something they could not have achieved independently (Provan and Kenis, 2008). This requires network-level competencies, which can relate to internal and external competencies. The former refers to coordinating skills and task-specific competencies, whereas the latter is about managing the external environment through, for instance, lobbying, acquiring funds, or building external legitimacy. Provan and Kenis (2008) argue that when the need for either type of competencies is high, a network tends to move from shared governance toward lead organization governance and later, NAO governance. Building collective leadership capacity is essential in collaboration (Stone, Crosby, and Bryson, 2013) and can be helpful in managing the need for these competencies. Research on interorganizational collaboration often emphasizes the role of facilitative leadership (Raišiene, 2012), as leadership can either facilitate or discourage cooperation among actors within an ecosystem (Emerson, Nabatchi, and Balogh, 2012).

Due to the interdependency among actors within a business ecosystem, organizational action is not only guided by organizational goals but by network-level goals as well. Actors within a business ecosystem should strive for general consensus on network-level goals. Provan and Kenis (2008) argue that organizational goals do not need to be similar but should overlap in terms of content and process. Shared governance is most suitable when there is general consensus (Provan and Kenis, 2008). Moreover, organizations need to agree on the resources to invest in the business ecosystem. A lack of commitment to the required resources, which can be any type of resources, like finance, time, physical space, materials, or personnel, impacts the effectiveness of collaboration

(Patel et al., 2012). This may emphasize the need for contractual governance mechanisms, focusing specifically on the control function.

Trust was discussed in section 2.1.1 as an element of relational governance. Previous relationships between players may enhance trust and commitment (Bryson et al., 2015) and therefore influence governance mechanisms and forms. The role of trust in interorganizational collaboration has been studied in different settings, treating trust either as an antecedent or moderator of relationship quality or as an outcome of collaboration. For instance, Lewicki and Bunker (1996) found that trust in professional relationships evolves gradually during different stages of trust development. As suggested by Vlaar et al. (2007), the presence of distrust increases the need for formal control (as part of contractual governance). Provan and Kenis (2008) refer to the distribution and reciprocity of trust among members of a network. When trust is widely distributed across actors within an ecosystem, there is a high level of trust. In contrast, when mutual trust is found only in dyads or small cliques, there is less trust. The level of trust may impact the governance form as well as the mechanisms employed.

As illustrated above, ambiguity remains with regard to how the governance forms and mechanisms evolve over time within a business ecosystem. Therefore, it seems promising to study the following research gaps:

> Investigate to what extent the development of governance mechanisms and governance forms is influenced by internal contingencies;
> To what extent is the development of governance mechanisms and governance forms influenced by:
> - the size of the business ecosystem?
> - the need for internal and external network competencies?
> - the extent of goal consensus within the business ecosystem?
> - the distribution of trust within the business ecosystem?

Conclusion

SMEs in all types of sectors – including agri-food – face the difficulty of having access to resources and specifically to knowledge, which is a major barrier for SMEs to spread technology. A network of partners within a business ecosystem allows SMEs to overcome these challenges, by relying on resources of partners within the ecosystem. Despite the opportunities that business ecosystems provide for SMEs, literature in this field remains incomplete. This chapter focused on two gaps in literature, which relate to the evolution of business ecosystems and business ecosystem governance. Based on a review of the literature, existing studies have been explored. Future research areas have been identified, which could lead to a better understanding of the business ecosystem construct.

References

Adner, R. (2017). Ecosystem as structure: an actionable construct for strategy. *Journal of Management*, 43(1), 39–58.

Adner, R., & Kapoor, R. (2010). Value creation in innovation ecosystems: how the structure of technological interdependence affects firm performance in new technology generations. *Strategic Management Journal*, 31(3), 306–333.

Ahuja, G., Soda, G., & Zaheer, A. (2012). The genesis and dynamics of organizational networks. *Organization Science*, 23(2), 434–448.

Barney, J. (1991). Firm resources and sustained competitive advantage. *Journal or Management*, 17(l), 99–120.

Battistella, C., Colucci, K., de Toni, A. F., & Nonino, F. (2013). Methodology of business ecosystems network analysis: a case study in telecom Italia future centre. *Technological Forecasting and Social Change*, 80(6), 1194–1210.

Bowen, F. E., Rostami, M., & Steel, P. (2010). Timing is everything: a meta-analysis of the relationships between organizational performance and innovation. *Journal of Business Research*, 63(11), 1179–1185.

Bryson, J. M., Crosby, B. C., & Stone, M. M. (2015). Designing and implementing cross-sector collaborations: needed and challenging. *Public Administration Review*, 75(5), 647–663.

Cao, Z., & Lumineau, F. (2015). Revisiting the interplay between contractual and relational governance: a qualitative and meta-analytic investigation. *Journal of Operations Management*, 33–34(1), 15–42.

Clarysse, B., Wright, M., Bruneel, J., & Mahajan, A. (2014). Creating value in ecosystems: crossing the chasm between knowledge and business ecosystems. *Research Policy*, 43(7), 1164–1176.

Delerue, H., & Simon, E. (2009). National cultural values and the perceived relational risks in biotechnology alliance relationships. *International Business Review*, 18(1), 14–25.

Deroïan, F. (2002). Formation of social networks and diffusion of innovations. *Research Policy*, 31(5), 835–846.

Dodgson, M., Gann, D. M., & Phillips, N. (Eds.). (2013). *The Oxford Handbook of Innovation Management*. Oxford: Oxford University Press.

Dolfsma, W., & van der Eijk, R. (2010). Knowledge development and coordination via market, hierarchy and gift exchange. In J. B. Davis (Ed.), *Global Social Economy: Development, Work and Policy* (pp. 58–78). London and New York: Routledge.

Donaldson, L. (2001). *The Contingency Theory of Organizations*. Thousand Oaks: SAGE Publications Inc.

Emerson, K., Nabatchi, T., & Balogh, S. (2012). An integrative framework for collaborative governance. *Journal of Public Administration Research and Theory*, 22(1), 1–29.

European Commission. (2018). Annual Report on European SMEs 2017/2018.

Fernández-Olmos, M., & Ramírez-Alesón, M. (2017). How internal and external factors influence the dynamics of SME technology collaboration networks over time. *Technovation*, 64, 16–27.

Gawer, A., & Cusumano, M. A. (2014). Industry platforms and ecosystem innovation. *Journal of Product Innovation Management*, 31(3), 417–433.

Griffith, D. A., Myers, M. B., & Harvey, M. G. (2006). An investigation of national culture's influence on relationship and knowledge resources in interorganizational relationships between Japan and the United States. *Journal of International Marketing*, 14(3), 1547–7215.

Iansiti, M., & Levien, R. (2004a). Strategy as ecology. *Harvard Business Review*, 82(3), 68–79.

Iansiti, M., & Levien, R. (2004b). *The Keystone Advantage: What the New Dynamics of Business Ecosystems Mean for Strategy, Innovation, and Sustainability*. Boston: Harvard Business Press.

Ikävalko, H., Turkama, P., & Smedlund, A. (2018). Value creation in the Internet of Things: mapping business models and ecosystem roles. *Technology Innovation Management Review*, 8(3), 5–15.

Institute on Governance. (2019). *Defining governance*. Retrieved from https://iog.ca/what-is-governance/

Jones, C., Hesterly, W. S., & Borgatti, S. P. (1997). A general theory of network governance: exchange conditions and social mechanisms. *The Academy of Management Review*, 22(4), 911–945.

Kilduff, M., & Brass, D. J. (2010). Organizational social network research: core ideas and key debates. *Academy of Management Annals*, 4(1), 317–357.

Kilduff, M., & Tsai, W. (2003). *Social Networks and Organizations*. London: Sage Publications Ltd.

Kraemer-Mbula, E., & Wamae, W. (2010). *Innovation and the Development Agenda*. Ottawa: OECD.

Larson, A. (1992). Network dyads in entrepreneurial settings: a study of the governance of exchange. *Administrative Science Quarterly*, 37(1), 76–104.

Lewicki, R. J., & Bunker, B. B. (1996). Developing and maintaining trust in work relationships. In R. M. Kramer & T. R. Tyler (Eds.), *Trust in organizations: Frontiers of theory and research*. Thousand Oaks: Sage Publications Inc.

Li, Y. R. (2009). The technological roadmap of Cisco's business ecosystem. *Technovation*, 29(1), 379–386.

Lu, C., Rong, K., You, J., & Shi, Y. (2014). Business ecosystem and stakeholders' role transformation: evidence from Chinese emerging electric vehicle industry. *Expert Systems with Applications*, 4, 14579–14595.

Moore, J. F. (1993). Predators and prey: a new ecology of competition. *Harvard Business Review*, 71(3), 75–86.

Moore, J. F. (1996). *The Death of Competition: Leadership and Strategy in the Era of Ecosystems*. New York: Harper Collins.

Moore, J. F. (1998). The rise of a new corporate form. *Washington Quarterly*, 21(1), 167–181.

Moretti, A. (2017). *The Network Organization: A Governance Perspective on Structure, Dynamics and Performance*. Cham: Palgrave Macmillan.

Ness, H., & Haugland, S. A. (2005). The evolution of governance mechanisms and negotiation strategies in fixed-duration interfirm relationships. *Journal of Business Research*, 58(9), 1226–1239.

Newig, J., Günther, D., & Pahl-Wostl, C. (2010). Synapses in the network: learning in governance networks in the context of environmental management. *Ecology and Society*, 15(4), 1–17.

Patel, H., Pettitt, M., & Wilson, J. R. (2012). Factors of collaborative working: a framework for a collaboration model. *Applied Ergonomics*, 43(1), 1–26.

Poppo, L., & Zenger, T. (2002). Do formal contracts and relational governance function as substitutes or complements? *Strategic Management Journal*, 23(8), 707–725.

Powell, W. (1990). Neither market nor hierarchy – network forms of organization. *Research in Organizational Behavior*, 12, 295–336.
Provan, K. G., & Kenis, P. (2008). Modes of network governance: structure, management, and effectiveness. *Journal of Public Administration Research and Theory*, 18(2), 229–252.
Provan, K. G., Fish, A., & Sydow, J. (2007). Interorganizational networks at the network level: a review of the empirical literature on whole networks. *Journal of Management*, 33(3), 479–516.
Raab, J., & Kenis, P. (2009). Heading toward a society of networks empirical developments and theoretical challenges. *Journal of Management Inquiry*, 18(3), 198–210.
Raišiene, A. G. (2012). Sustainable development of inter-organizational relationships and social innovations. *Journal of Security and Sustainability Issues*, 2(1), 65–76.
Reuver, M. de, & Bouwman, H. (2012). Governance mechanisms for mobile service innovation in value networks. *Journal of Business Research*, 65(3), 347–354.
Roehrich, J. K., Selviaridis, K., Kalra, J., Van Der Valk, W., & Fang, F. (2019). Inter-organizational governance: a review, conceptualisation and extension. *Production Planning & Control*, 1–17.
Rogers, E. M. (1995). *Diffusion of Innovations* (4th ed.) New York: The Free Press.
Rong, K., Hu, G., Lin, Y., Shi, Y., & Guo, L. (2015). Understanding business ecosystem using a 6C framework in Internet-of-Things-based sectors. *International Journal of Production Economics*, 159(SI), 41–55.
Rosenbusch, N., Brinckmann, J., Bausch, A., & Bausch, A. (2011). Is innovation always beneficial? A meta-analysis of the relationship between innovation and performance in SMEs. *Journal of Business Venturing*, 26, 441–457.
Rousseau, D. M., Sitkin, S. B., Burt, R. S., & Camerer, C. (1998). Not so different after all: a cross-discipline view of trust. *Academy of Management Review*, 23(3), 393–404.
Scaringella, L., & Radziwon, A. (2018). Innovation, entrepreneurial, knowledge, and business ecosystems: old wine in new bottles? *Technological Forecasting and Social Change*, 136(1), 59–87.
Stone, M. M., Crosby, B. C., & Bryson, J. M. (2013). Adaptive Governance in Collaborations. In C. Cornforth & W. A. Brown (Eds.), *Nonprofit Governance: Innovative Perspective and Approaches*. Abingdon: Routledge.
Thompson, J. D. (2003). *Organizations in Action: Social Science Bases of Administrative Theory*. New Brunswick: Transaction Publishers.
Turber, S., Vom Brocke, J., Gassmann, O., & Fleisch, E. (2014). Designing Business Models in the Era of Internet of Things: Towards a Reference Framework. *International Conference on Design Science Research in Information Systems*, 8463, 17–31.
Van Burg, E., Podoynitsyna, K., Beck, L., & Lommelen, T. (2012). Directive deficiencies: how resource constraints direct opportunity identification in SMEs. *Journal of Product Innovation Management*, 29(6), 1000–1011.
Vangen, S., Hayes, J. P., & Cornforth, C. (2015). Governing cross-sector, inter-organizational collaborations. *Public Management Review*, 17(9), 1237–1260.
Vlaar, P. W. L., Van Den Bosch, F. A. J., & Volberda, H. W. (2007). On the evolution of trust, distrust, and formal coordination and control in interorganizational relationships toward an integrative framework. *Group & Organization Management*, 32(4), 407–429.
Wang, E. T. G., & Wei, H.-L. (2007). Interorganizational governance value creation: coordinating for information visibility and flexibility in supply chains. *Decision Sciences*, 38(4), 647–674.
Wareham, J., Fox, P. B., & Cano Giner, J. L. (2014). Technology ecosystem governance. *Organization Science*, 25(4), 1195–1215.

Westerlund, M., Leminen, S., & Rajahonka, M. (2014). Designing business models for the Internet of Things. *Technology Innovation Management Review*, 4(7), 5–14.

Williamson, O. E. (1991). Comparative economic organization: the analysis of discrete structural alternatives. *Administrative Science Quarterly*, 36(2), 269–296.

Young, H. P. (2009). Innovation diffusion in heterogeneous populations: contagion, social influence, and social learning. *American Economic Review*, 99(5), 1899–1924.

Zaheer, A., & Venkatraman, N. (1995). Relational governance as an interorganizational strategy: an empirical test of the role of trust in economic exchange. *Strategic Management Journal*, 16(5), 373–392.

Zaheer, A., Gözübüyük, R., & Milanov, H. (2010). It's the connections: the network perspective in interorganizational research. *Academy of Management Perspectives*, 24(1), 62–77.

Zahra, S. A., Neubaum, D. O., & Naldi, L. (2007). The effects of ownership and governance on SMEs' international knowledge-based resources. *Small Business Economics*, 29(3), 309–327.

7 The social impact of ICT-enabled interventions among rural Indian farmers as seen through eKutir's VeggieLite intervention

Spencer Moore, Maria Carmela Annosi, Teun Gilissen, Jennifer Mandelbaum, and Laurette Dube

Introduction

The agriculture–food value chain is at the core of economic growth and human development in India. Agriculture and food sectors contribute over 20% to India's GDP and employ more than 50% of India's population. At the same time, the agri-food sectors also shoulder the onus of securing the nutritional security for a population that is simultaneously fighting poverty, undernutrition, and overnutrition (Moore et al., 2015). Various approaches – for and not-for-profit – have been offered as a means of addressing the endemic nature of rural poverty and undernutrition. eKutir, a social enterprise, has sought to address rural poverty through a team of microentrepreneurs that leverage information and communication technologies (ICT) to connect farmers more efficiently with markets. Microentrepreneurs engage with farmers on an everyday basis using eKutir's ICT-based digital platform (Jha et al., 2016). Microentrepreneurs are trained in (social) entrepreneurship, ICT, inputs, and market access and best practices in farming, including farm soil-testing, fertilizers, seeds and crop nutrients, and pest management. Each microentrepreneur manages roughly 200 smallholder farmers.

Core to eKutir's ICT platform is eAgrosuite, which consists of various digital technologies capturing, monitoring, and managing data and various transactions. For example, a nutrition management tool called mrittika (soil in local language) was developed to test soil quality both quickly and affordably and provide recommendations to farmers about what types of and how much fertilizer to use. Another digital tool called ankur provides recommendations on which seed varieties to use based on local soil and climate conditions, the type of crop being sought, and the growing season. Additionally, a Farmer Portfolio Management Tool, which links farmers to agricultural experts, agricultural input providers and buyers, was developed to streamline farmers' production and distribution channels. The data gathered from individual farmer transactions are used to create a more efficient and effective supply of agricultural produce to the market, obtain high quality inputs at a lower price for farmers, and a better sales price for the farmers (Jha et al., 2016).

Social impact

Through its local microentrepreneurs and ICT platform, eKutir has also organized its users into groups of 15–25 members called farmer intervention groups (FIGs). These communities of practice consist of microentrepreneurs and eKutir farmers sharing information and best practices within local FIGs (Moore et al., 2015). FIG members do not yet use digital technology while interacting, but ICT is used to form and manage the FIG by optimizing communication and making offline meetings and interactions increasingly possible in the future. In the process, eKutir has thus fostered a social ecosystem through which it introduces its ICT platform to rural farming communities and engages with local farmers. eKutir's ecosystem serves in practice to re-position farmers within their local social network, giving them access to quality information sooner than farmers not participating in the eKutir program.

A social broker is someone who acts as intermediary between two or more social actors (Burt, 2005). In social network terms, social brokerage represents the capacity that a person may have in coordinating actions across otherwise disconnected parts of a network. Social brokers thus occupy bridging positions within social networks and enable the flow of information and resources more widely across a network (Burt, 2015). Farmers having higher social brokerage have earlier access to diverse market information and an advantage in detecting market opportunities (Burt, 2005). In the remainder of this chapter, we will discuss how eKutir's ecosystem acts on the social brokerage position of its farmer members.

VeggieLite intervention

In 2014, eKutir received funding from Grand Challenges India and the Biotechnology Industry Research Assistance Council of India to implement and evaluate its VeggieLite program among farmers and consumers in Odisha, India. VeggieLite was a value-chain level intervention that aimed to address gaps in the affordability and availability of vegetables and fruits among low-income rural and urban communities. Through the application of the eKutir microentrepreneurial model in rural Odisha and the additional creation of retail microentrepreneurial services, the intervention sought to raise the productivity of farmers, farmer's income, and intake of fruits and vegetables, and make fruits and vegetables more available and affordable to individuals and households residing in slum areas of Bhubaneswar, India. Further details about the purpose of VeggieLite intervention can be found elsewhere.

To assess the effectiveness of the VeggieLite intervention, we used a quasi-experimental evaluation design, with pre- and post-intervention data collected one year apart. To evaluate the effects of the intervention, a rural and urban study sample was created. For the purposes of this chapter, we focus exclusively

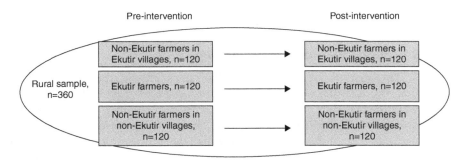

Figure 7.1 VeggieLite evaluation framework in rural Odisha.

on the rural arm of the VeggieLite intervention. Figure 7.1 illustrates the overall evaluation design for the rural arm of the intervention.

Among the 32 rural villages taking part in the evaluation, our quasi-experimental design consisted of three groups: (1) rural farmers using the agricultural microentrepreneurial services and having access to eKutir's retail markets (FMEs+RMEs), (2) rural farmers with access only to the retail markets (RMEs), and (3) farmers in villages unexposed to the eKutir ecosystem (comparison farmers). This three-group design was meant to enable a series of comparative assessments of the effect of the eKutir ecosystem on fruit and vegetable consumption, purchase and production patterns of rural farming households. Further details on the evaluation design may be found elsewhere (Dubé et al., 2020).

Structured household questionnaires were administered at pre- and post-intervention to participating head of households, with questions about household demographics, agricultural production and cropping, fruit and vegetable consumption, and social networks. Baseline questionnaires were administered to the heads of rural households in April/May 2015 with endline questionnaires administered to rural households in April/May 2016 (Moore et al., 2015). For the purposes of this chapter, we focus on the social network module of the rural questionnaires.

The social network module consisted of two name-generator questions asking farmers to name up to three other farmers with whom they have discussed about agriculture in the last three months and up to three farmers with whom they might discuss food matters in the last three months. Farmers might thus name anywhere between zero and six farmers with whom they had agricultural or food discussions. We reviewed the list of farmers' names separately for each of the 32 villages in the VeggieLite intervention to remove redundant names and create a matrix indicating the connections among farmers in the study villages. This matrix was used to construct a quasi-whole network map of each study village, with these maps used to calculate the social brokerage position of the farmers participating in the evaluation study.

Social brokerage among rural farmers

Figure 7.2 is a snapshot of one part of a village network with the red square or node indicating a farmer with a higher level of social brokerage in the network.

To examine whether the eKutir social ecosystem might affect the social brokerage position of eKutir-participating farmers, we measured the social brokerage of farmers at the beginning and end of the intervention. Furthermore, to assess whether the intervention might affect the social brokerage of eKutir-participating farmers differently than non-participating farmers, we compared the scores of eKutir-participating farmers with those of the other quasi-experimental groups (i.e., the RMEs and comparison farmers).

Multilevel linear regression was used to assess the degree to which the endline levels of social brokerage differed among the farmer groups, while adjusting for baseline levels of social brokerage and accounting for the clustering of farmers

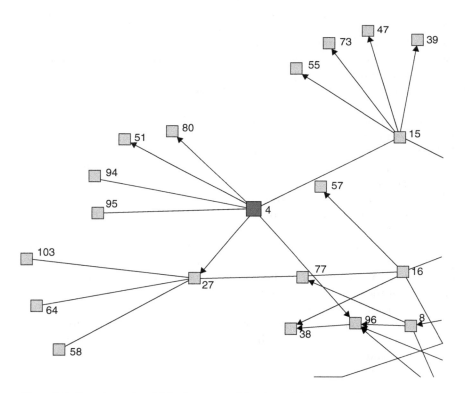

Figure 7.2 Snapshot of social brokerage position in a village network.

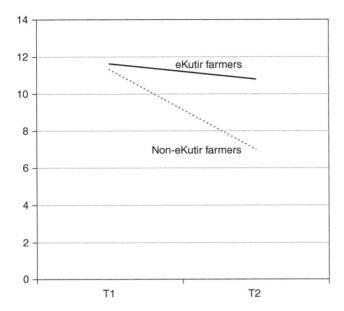

Figure 7.3 Graph comparing one-year change in social brokerage between eKutir-participating farmers and non-participating farmers.

within village networks. Figure 7.3 illustrates graphically the results of these statistical analyses.

Analyses show that the brokerage position of non-participating eKutir farmers declined significantly in comparison to eKutir-participating farmers over the course of one-year intervention. The findings thus suggest that the eKutir ecosystem may have acted to buffer or protect its participating farmers against fluctuations in their brokerage position within villages. In contrast, those farmers outside the eKutir ecosystem of programs – ICT, microentrepreneurs, and FIGs – were more vulnerable to changes in their network connectivity and brokerage role within villages.

Conclusion

Within the context of an ICT-enabled intervention among rural farmers, our study highlights the social impact that such interventions may have on the social networks of rural farmers. Social network data taken from the evaluation of eKutir's VeggieLite intervention suggest that the broader social ecosystem that eKutir created around its microentrepreneur-based program operated in a protective fashion to preserve the social brokerage position of its farmer members. Further research is needed to assess whether this protective role was due in greater part to the ICT-enabled parts of the intervention or the FIG activities.

In addition, comparative research is needed to assess whether interventions led by other social enterprises have similar social impacts on their participating members.

References

Burt, R. (2005). *Brokerage and Closure*. Oxford: Oxford University Press.

Dubé, L., McRae, C., Wu, Y., Ghosh, S., Allen, S., Ross, D., Ray, S., Joshi, P., McDermott, J., Jha, S., Moore, S. (2020). Impact of the eKutir ICT-enabled social enterprise and its distributed micro-entrepreneur strategy on fruit and vegetable consumption: a quasi-experimental study in rural and urban communities in Odisha, India. *Food Policy*, India, 90, 101787.

Jha, S. K., Pinsonneault, A., Dubé, L. (2016). The evolution of an ICT platform-enabled ecosystem for poverty alleviation: the case of Ekutir. *MIS Quarterly*, 40(2), 431–445.

Moore, S., Jha, S., Mishra, S., Ross, D., Allen, S., Dubé, L. (2015). Innovation in Evaluation to Inform Policy Convergence: Complex Systems Approach to Assess Entrepreneurship-driven Intervention in eKutir's VeggieKart, pp. 197–208, in *Evaluations for Sustainable Development: Experiences and Learning*, New Delhi: Daya Publishing House.

Part IV
The evolution of digitalization in agri-food

8 Artificial intelligence

Toward a new economic paradigm in agri-food

Alessio Gosciu, Federica Brunetta, and Maria Carmela Annosi

Introduction

Artificial intelligence (AI) is finally providing a multitude of competencies to machines which were long thought to belong exclusively to humans. As mentioned in the previous chapters, today, intelligent computers can process natural language or visual information, making decisions and interacting with humans and external environments autonomously. The range of machines' capabilities is continuously expanding while their artificial minds become ever more like the human brain. The implication is that the diffusion of AI is driving work automation and is changing the way organizations operate.

While the concept of AI dates back to more than 50 years, only recently have technological advances enabled successful development and implementation at the industrial scale. The first person to use the term Artificial Intelligence was John McCarthy in 1956 during the first academic conference on this topic: the Dartmouth Conference (Childs, 2011). However, the first idea of AI appeared a decade before with Bush (1945), which imagined a "system" able to amplify individual knowledge and understanding. In 1950, Alan Turing was the first to write about "computing machinery" and "digital computers" able to imitate human beings, both in acting and in cognitive skills (Bringsjord et al., 2003).

More generally, AI encompasses a set of technologies that provide computers with cognitive skills, such as problem-solving, decision making, and reasoning, as well as voice, image, and language recognition. In sum, AI is about providing *machines* with *intelligence*. After decades of research and development, the progress and the importance of AI are flourishing. This is related to advancements that allow the application of AI across a variety of applied domains (Calo, 2017), such as: (i) the expanded availability of data; (ii) increase of computing power; (iii) more powerful algorithms; and (iv) decreasing storage costs. In this light, the list of AI applications is expanding and includes, among others: (i) robotics and autonomous vehicles; (ii) computer vision, (iii) natural language processing (NPL); (iv) virtual agents; and (v) machine learning.

Nonetheless, the disruption caused by AI does not only depend on the technology itself but also on the reaction of managers, workforces, and organizations

(Kolbjørnsrud et al., 2016) and on the policy responses authorities and institutions are able to offer (Calo, 2017). AI is impacting all industries and is requiring companies to rapidly readapt as a strategic competitive advantage, reflexing on the firms' growing revenues (Ransbotham et al., 2019).

In this fastly growing environment, institutions are called upon to redesign boundaries and incentives in order to alleviate biases and unlock the potential of such innovation. AI-driven innovation will continue to create wealth and foster economic growth. Additionally, with the implementation of new technologies, new skills are needed for workers and new strategies are indispensable (Kolbjørnsrud et al., 2016).

Nowadays, the diffusion of the Internet is blurring differences between the traditional economic agents: consumers can also be producers, firms start to play the same role as national institutions, and national borders are vanishing. Business model are challenged, with AI casting the economy into a transformation stage between a traditional economic structure and the new economic paradigm.

Main applications of AI

Organizations are increasingly perceiving AI not only as an opportunity, but also as a competitive risk, such as the one arising from competitors that, by implementing AI, becomes more efficient in reducing costs and/or creating value, exacerbating existing competition, or new entrants that disrupt related – or even unrelated – industries creating new threats (Ransbotham et al., 2019). With managers perceiving AI as both a risk and opportunity, the urge to build AI competencies is on the verge. Companies exploit AI for both cost-cutting and value-increasing activities, but most especially, they are taking up opportunities that generate revenue.

AI enhances the performance of assets and optimizes the process of transforming input into output while ensuring reliability and quality; all this translates into enhanced productivity. There are many examples of AI-based applications in production process. The convergence between human and AI automation is a powerful tool, especially if such convergence is directed toward speeding up the performance of production process while at the same time enhancing quality and reducing errors.

Additionally, AI has a crucial role in projecting and forecasting trends and patterns using the analysis of vast flows of data. Sophisticated algorithms extract information and can recognize paths and trends in the behavior of demand. The knowledge of future movements is a powerful competitive advantage because it has a huge impact on costs and the enhancement of assets. Predicting the future behavior of customers is crucial in shaping marketing strategies, anticipating sale trends, and better managing the supply chain. Thus, the main benefits of AI-based approaches impact supply chain management, research, development operations, and functions supporting businesses.

In business, AI-enhanced supply chain management increases the precision of forecasts and optimizes asset utilization. For example, the vast amount of data from sensors and databases are absorbed by machine-learning applications which are today able to estimate the useful life of devices and accurately detect the time of decommissioning (McKinsey global institute, 2017). Traditionally, the expiration time of an asset has been estimated, but with AI, enterprises are able to identify the precise moment of its end. Indeed, the efficiency of the supply chain is crucial for the success of a company, and the key strategy to achieve a competitive advantage is creating the perfect combination of demand prediction and inventory replenishment strategies. Machine-learning technology is a key tool for powering supply chain optimization, with reinforcement learning systems as the most advanced implementation of AI for inventory management. They not only allow machines to produce forecasts but to directly act on them without human intervention.

Finally, AI is a powerful tool for promotion and personalizing experience. In the retail industry, AI performs tasks such as selecting the amount of a discount and personalizing promotion in relation to information about the customer, like age, gender, habits, and much more. AI enhances the precision of the offer and its efficacy because algorithms know what the shoppers want before the shopper themselves.

How is AI supporting agri-food?

Organizations operating within the agri-food industry are subject to pressure related to the need of reducing costs and/or creating value, not only for competitive reasons but also for the scrutiny related to sustainability challenges that will require them to support the needs of a growing world population (FAO, 2013), with a subsequent need to increase production by 70% (FAO, 2012).

In order to do so, innovation will be key (Morand and Barzman, 2006) and AI is a necessary element of the industry's technological evolution (Kosior, 2018). Thus, any delay in adopting technology and innovation exposes companies not only to increased threats of competition but also to societal imperatives related to environmental challenges (Long et al., 2016).

Within the agricultural industry (Kosior, 2018), AI supports the development of: (i) agricultural robots handling tasks such as seeding and harvesting at higher rates; (ii) crop and soil monitoring, by leveraging computer vision, big data analysis, and machine learning in combination with hardware solutions (such as drones); and (iii) predictive analytics, through algorithms that project and forecast trends and patterns through the analysis of vast flows of data to detect potential issues on crop seed, yield, and harvest. Indeed, farms that are investing in technology are also on track to produce a vast amount of data, monitoring in real time soil, water, temperature, and weather. Through precision agriculture, the use of AI technology can be used to predict and forecast, which is invaluable to most farmers as well as small and medium-sized farms, which alone carry out 70% of the production. Additionally, it is not uncommon for AI tools to be used for assistance, such as chatbots (Kosior, 2018).

Moving forward along the supply chain, the use of AI within the food industry is multifaceted, such as being applied to production, manufacturing, and commerce. Early adopters have already applied AI to various tasks within their operations, starting from those areas requiring the optimization and processing of large volumes of data, as well as highly routinized tasks, allowing for a shift of human resources toward different strategic functions.

Indeed, food supply chains are often multi-sided, with different stakeholders covering different activities, objectives, and priorities. The coordination among such an intricate group might be too complex to be managed without coordination costs. Nonetheless, these could be significantly moderated using AI, which can constantly monitor, process, and control all the different activities, leveraging diverse technologies, such as the Internet of Things. The same can be applied to the transportation of food, since the perishability of specific food products can be prevented by an efficient transportation management system. Maintenance is another important field of AI application within the food industry. The optimization of asset utilization and the predictive capability of AI through data collected from sensors and databases are crucial: automated predictive maintenance is used to monitor equipment status as well as predict and plan maintenance activities, ensuring continuity of operations, use of resources, and process efficiency. A more efficient supply chain results in less waste, faster time to market, and better product quality. Another powerful effect is the development of automated food processing, which ensures complete hygiene and high food quality and safety.

Finally, retail and food delivery benefit tremendously from AI, which can be seen as an enabler of new business models and marketing efforts by monitoring individual customer behavior, ordering and delivery, and customer service, such as through AI-powered chatbots.

Conclusion

The transformation of the agri-food sector in the light of the spread of AI is undeniable. However, and despite the proven efficacy of such technologies, we are only looking at still limited and initial applications of AI technology. Today, we are still not able to evaluate the indirect future effects of intelligent technologies with enough reliability due to this early stage of AI development and the uncertainty of related spill-over effects, effects that may be positive but may also create critical challenges that require new tools and innovative approaches. As these AI systems advance and become more accessible, their spread likely occurs at a faster pace. Stephen Hawking said that "Artificial Intelligence could be the biggest event in the history of our civilization. Or the worst. We just don't know." The outcome depends on the capacity of our economy to adapt to the changes, and we need new tools. The stakes are high, but so are the opportunities.

While we cannot try to prevent innovation from occurring, we must learn how to manage it. We begin by providing two examples of virtuous companies that have successfully embraced the "AI revolution." In the following chapter, we present two cases of AI diffusion and application in the agriculture and food industry, respectively.

References

Barzman, M., & Morand, F. (2006). European sustainable development policy (1972–2005): fostering a two-dimensional integration for more effective institutions. *Working Paper* retrieved at: https://hal.archives-ouvertes.fr/hal-01189947

Bringsjord, S., Bello, P., & Ferrucci, D. (2003). Creativity, the Turing test, and the (better) Lovelace test. In *The Turing Test* (pp. 215–239). Springer, Dordrecht.

Bush, V. (1945). As we may think. *Resonance*, 5(11).

Calo, R. (2017). Artificial intelligence policy: a primer and roadmap. *UCDL Review*, 51, 399.

Childs, M. (2011). John McCarthy: computer scientist known as the father of AI. *The Independent Digital News and Media*, 1.

FAO (2012). *World agriculture towards 2030/2050: the 2012 revision*. In: FAO (Ed.). Food and Agricultural Organisation of the United Nations, Rome, Italy.

FAO (2013). Climate-smart agriculture – sourcebook in nations. In: FAO (Ed.). *Food and Agriculture Organisation of the United Nations*, Rome, Italy.

Kolbjørnsrud, V., Amico, R., & Thomas, R. J. (2016). How artificial intelligence will redefine management. *Harvard Business Review*, 2–6.

Kosior, K. (2018). Digital transformation in the agrifood sector – opportunities and challenges. *Roczniki* (Annals), XX(2), 98–104.

Long, T. B., Blok, V., & Coninx, I. (2016). Barriers to the adoption and diffusion of technological innovations for climate-smart agriculture in Europe: evidence from the Netherlands, France, Switzerland and Italy. *Journal of Cleaner Production*, 112, 9–21.

McKinsey Global Institute (2017). *Artificial Intelligence, the Next Digital Frontier?* Retrieved at: www.mckinsey.com/mgi/overview

Morand, F., & Barzman, M. (2006). European sustainable development policy (1972–2005): fostering a two-dimensional integration for more effective institutions. Hal Archives-ouverts.fr. Retrieved at: www.hal.archives-ouvertes.fr/hal-01189947

Ransbotham, S., Khodabandeh, S., Fehling, R., LaFountain, B., & Kiron, D. (2019). Winning With AI. *Sloan Management Review*, 10.

9 Agri-food and AI

Integrating technology as a core element of the business model: the case of Elaisian

Alessio Gosciu and Federica Brunetta

Introduction

Elaisian is a precision farming service focused on the preservation of olive trees. Thanks to a system of algorithms based on a database of agronomic studies and machine learning techniques, the company is able to prevent diseases and optimize cultivation processes, such as irrigation and fertilization. With its technology, olive oil producers are able to constantly monitor the situation in the field through sensors installed in the ground. These devices collect climatological data, chlorophyll, and soil components. The algorithms process information and through machine learning and other databases provide real-time information on the status of trees and suggest advice. Farmers receive science-based suggestions on optimizing irrigation and the efficient use of fertilizers. Elaisian's devices gather climatological data such as rainfall, humidity, and temperature and analyze chlorophyll and beta-carotene levels. At the same time, the intelligent software pre-warns producers about diseases and enables treatment when diseases attack. Elaisian utilizes AI to elaborate and predict the behavior of the land by providing increasingly precise insights. The algorithm learns exactly when, where, and for how long to irrigate, cutting costs and waste. The brain is the company platform where all the data and information are transmitted through Wi-Fi. An algorithm processes the data collected on the ground and cross-references it with other sources, such as historic databases, weather information, satellite images, and other agronomic data. The purpose is to improve the life of the olive trees and the producer's performance. Intelligent technologies maximize production and enhance cultivation processes.

Every year, on average, 60 percent of olive oil production is lost due to inefficient tools. According to Elaisian, their operative costs are reduced by 20 to 30 percent while production is enhanced up to 20 percent. Thanks to water monitoring, carbondioxide emissions in the production process are cut by more than 18 percent. Seventy percent of global water consumption is used in agriculture, of which 50 percent is wasted. Irrigation expenses have a fundamental impact on operative costs. Intelligent algorithms are able to lower these costs and increase efficiency in water consumption.

Elaisian was established in November 2016 based on an idea by its two co-founders. After three months of development, the start-up was officially launched under the guidance of a renowned acceleration program. In terms of economics, the program made an initial investment of €15,000 and 6 percent of the company's equity. This investment enabled the implementation of the first minimum viable product and the creation of a commercial partnership, as well as a consultation from top managers in the start-up ecosystem and European–Italian food industry. After three months of this, Elaisian attended the "Demo Day" event in front of an audience of more than 300 investors. The event raised additional cash and investment, which enabled further improvements and a commercial development of the product.

Today the team comprises six people: the two founders (CEO and CMO), one agronomic expert, two IT engineers, and one salesman. In July 2018, new funds were provided through an online crowdfunding campaign. Elaisian now operates in several regions with 50 olive oil farms and 80,000 olive trees monitored.

Market analysis

Elaisian proposes an agricultural business intelligence solution founded on in-field measurements. The company operates in the food tech industry providing business-to-business service in the agriculture pre-harvesting production phase. In particular, Elaisian is specialized in olive oil production.

In general, the reference field is the agriculture tech sector, which aims to increase farm efficiency through software, sensors, aerial-based data, internet-based distribution channels (marketplaces), and tools for technology-enabled farming. Interests and expectations are growing in this market since the acquisition of "Climate Corporation" by Monsanto for more than $1billion in 2013. Climate Corporation was one of the first to move in the collection of field data for farmers in order to enhance productivity and optimize costs. The billion-dollar acquisition increased attention on the sector.

In 2017, the amount of funds and investments in AgTech nearly doubled, surpassing $700 million globally. In the same year, in the sector, there were more than 300 unique investors and more than 160 deals, compared with the 31 deals and less than $200 million in investment in 2007 (CB Insights, 2018).

Corporate venture capital has raised interest in the sector. Indeed, there are now more than 30 active funds, plus AgTech-focused funds like Khosla, Fall Line, Finistere, Innovation Endeavors, and S2G. In this landscape, data-driven agronomy implemented with artificial intelligent tools is the new market trend. The implementation of on-field sensors and artificial intelligence platforms are driving the evolution from precision to "predictive" agriculture (Kukutai and Maughan, 2018).

However, Elaisian also depends on the olive oil market. Italy represents a profitable market: there are 3.5 million oil producers worldwide, with 850,000 of them in Italy alone, hobbyists included. Seventy-five percent of them comprise small firms with up to five employees. Elaisian's targets are only the firms that constitute roughly 25 to 35 percent of the entire market (Wired and IBM, 2014).

Competition analysis

The broader competitive landscape is dominated by the two largest players, Dow/DuPont and Bayer/Monsanto, which control varied segments in the market. At the same time, non-traditional new entrances are approaching the sector through partnerships and M&A. Most of these are digital giants that are exploiting core competences, such as Google for data, Amazon for its supply chain, and Facebook for connecting and selling to farmers.

With concern to Elaisian, the strategic market of reference is the in-field measurements through sensors. This highly competitive *sector* vaunts a large number of start-ups. However, Elaisian is the only existing start-up specialized in olive oil. Furthermore, the geographic distribution of olive trees determines the range.

The reference sector is the widest AgTech but Elaisian, in particular, is active in precision farming, sensors, and data analytic segment.

Elaisian can count on AI expertise provided directly by a partnership with a local university. This is one of the most relevant strengths because the relationship between company, university, and local farms creates a favorable ecosystem. This is enhanced by a strong geographical permeation in Italy, one of the top global producers of olive oil. Furthermore, the geographic concentration of olive oil production can be a positive factor for company logistics; 80 percent of olive trees are in southern regions. Focusing on a smaller specific area may enable economies of scale and scope for what concerns data collected in the territory. This strength has been pursued over the last year with the expansion into the Spanish market. Spain is the largest global olive oil producer and consumer; its farms harvest almost the half the world's production. Elaisian can take advantage of its brand, trustworthy and specialized on olive oil. This allows the company to build more precise and specific data for a single objective.

This strength may be useful in facing the increasing competitiveness of the market. Since higher volumes of investments are flowing in AgTech, the interest of biggest players both inside and outside the industry is growing. Numerous start-ups are growing that are specialized in in-field measurements, and many of these are backed by large venture capital investments or multinational tech giants. Elaisian is not the only firm in this sector but it is the only and the first to specialize in olive oil. While there are still no big competitors in the Mediterranean area, there are dozens of new companies in the US market. The most concentrated area is California, home to Silicon Valley and where the weather is similar to the Mediterranean climate. As such, there is a possibility for existing and new companies to specialize in olive oil.

What can be a threat can also be an opportunity. If big investors are willing in the future to permeate the Mediterranean area, they could also decide to acquire

Elaisian and offer a rich exit strategy. However, the presence of big investors and the lack of AI expertise may raise the price of few technicians who are able to run intelligent and complex systems. In addition, another threat that may also be an opportunity is the spread of olive trees illnesses. Elaisian's business is currently related to the performance of olive oil producers. While the fear of tree illnesses is a strong incentive for farmers to adopt a control system, if the disease becomes uncontrollable, it would devastate the entire sector. However, over the last decade, there has been a growing interest in high-quality olive oil, even more so than wine. This is a good sign for Elaisian given its dependency on this sector. Furthermore, rising ecological concerns are prompting the need for sustainable food options that have a lower impact on the environment.

The growing population is driving the surge in demand. Combined with the challenges posed by climate change, this creates a severe urgency to increase yield efficiency and optimize the exploitation of natural resources like water. Climate change can be both an opportunity and a threat that increases the complexity of efficient field management.

Technological readiness

Measuring the technological readiness of the farm industry means detecting the rate of resistance to new technological tools and innovation. If one industry is already digitalized, it will be easier to launch newly updated tools. This is particularly significant for companies like Elaisian and for innovations such as intelligent machines.

The production process is the step-in value chain in which Elaisian provides its services. The data suggest a high attention on the part of farmers on the efficient utilization of natural resources and reduction of wastes which impact operative costs more. The same research indicates that the leading technology is the utilization of drones for monitoring fields, while the second most appreciated is "sensomining," in-field sensors for precision agriculture. However, while such technologies are widely valued by farmers, their dissemination is still low. Almost 70 percent of respondents are not willing to implement new technologies in the short run. It emerges, though, that the most diffused innovation are sensors and big data analysis.

According to Elaisian's research, olive oil farming still has a low digitalization rate, around one percent.

Artificial intelligence and the Elaisian case

The aim of the analysis of this company is to verify the positive impact of intelligent technologies on a concrete business case. The decision to choose this firm is related to the elevated utilization of AI-related technologies in the business value chain. In particular, Elaisian has incorporated into its core business AI applications such as machine learning technologies through which algorithms learn how to process data and provide more precise predictive analysis.

Input data are provided through external sources, such as satellite images and historic data on the geographic area concerning weather trends. Furthermore, farm-specific data come from in-field collection through a device installed directly in the ground. The device is a sort of computer that covers up to six hectares of field. Data are transmitted to the platform that processes it through algorithms based on agronomic studies. The farmer has his own access to the company web platform, where he receives all the necessary information. Advice regarding daily cultivation and alarm notifications for any diseases are provided based on real-time observation.

There are two Elaisian AI applications: Machine Learning and Predictive Analytics applied through algorithms to the huge amount of data stored in the company database. The reason why Elaisian is interested in adopting such technologies is because the company "system is all based on data collection and processing. In this sense, both machine learning and predictive analytics are fundamental." How important are artificial intelligent applications for Elaisian's business? "This technology allows Elaisian to have a self-updating and more precise algorithms with reduction of personnel costs and maintenance. (…) Through machine learning implementation all processes are automated". Elaisian hopes to achieve new ways to enhance automation of process and its digitalization. At the same time, AI is a precious asset in assisting and implementing new business strategies.

> While it is hard to find experts because there are few of them and they are contented by big tech giants, we aim to internally develop our expertise. Two reasons: it is cheaper, and we can have more control over it since it can constitute a competitive advantage.

Founders believe that new intelligent applications are more impactful. This is a crucial issue, since Europe is lagging behind China and the US for AI firm implementation; Elaisian is ready to take up the challenge.

References

CB Insights (2018) *CB Insights from Ag tech industry*, 2018. Retrieved from: www.cbinsights.com/research/ag-tech/

Kukutai, A., & Maughan, S. (2018) Major trends in Agtech for 2018. TEchcrunch. Retrieved from: https://techcrunch.com/2018/03/08/major-trends-in-agtech-for-2018/?guccounter=1

Wired and IBM (2014). Agrinnova. Retrieved from: www.wired.it/partner/agrinnova/

10 Convergence research and innovation digital backbone

Behavioral analytics, artificial intelligence, and digital technologies as bridges between biological, social, and agri-food systems

Laurette Dube, Sjaak Wolfert, Karin Zimmerman, Nathan Yang, Fernando Diaz-Lopez, Rigas Arvanitis, Sandra Schillo, Sabina Hamalova, Jian Yun Nie, and Shawn Brown

Introduction

Food is a powerful bridge between human biology and the agro-ecological, social, cultural, and economic contexts in which we live, be it in traditional or modern economies and societies. This bridging power, however, is in clear need of evolution as diet-related challenges to human and planetary health are among the most pressing societal issues (James et al., 2004). For instance, in Western countries like Canada, where consumers are offered an ever increasing diversity of appealing food products and brands of variable nutrition quality, over 60% of the adult population is overweight or obese (Rodd & Sharma, 2016). Excess body weight is a significant risk factor for several diseases including heart diseases and cancer, which are the top causes of death in North America (Must et al., 1999). Alarmingly, the life expectancy in the United States has decreased for two years in a row, in part due to diet-related chronic diseases and overweight/obesity (Kochanek et al., 2017). Even a country like India, which has high rates of undernutrition and where subsistence agriculture is progressively giving way to agri-food supply chains and markets, has high rates of diabetes, a chronic disease largely tied to overnutrition (Subramanian et al., 2007). These epidemiological facts highlight the urgency for more effective strategies to support both individuals in their struggle for lifelong nutrition, as well as those actors throughout economy and society that define contexts for food choices and behaviors.

These strategies need to account for the complexity tied to healthy diets from sustainable systems consisting of food at the converging point, i.e., food that (1) consumers want; (2) they need for their vitality and health; (3) they are able and willing to pay for; (4) the planet can offer in a sustainable way; and (5) the actors in the agriculture and food sectors can and want to produce in a cost-effective and economically viable manner (Dube et al., 2012, 2014a). This requires better convergence between economic, social, health, and environmental processes and outcomes guiding decisions at all levels in the agri-food system: individual, professional, organizational, system, and policy (Dube et al., 2012, 2014a). This is being called for by the United Nations in setting goals for integrative sustainable development (United Nations, 2017; Weitz et al., 2014) as well as by leading research and infrastructure agencies now placing convergence among a few top big ideas for a 21st century science.

For instance, the 2018–2023 strategic roadmap of the Canada Fund for Innovation (CFI) states that to address 21st-century grand challenges

> requires the deep integration of disciplines, knowledge, theories, methods, data and communities. Convergence goes beyond interdisciplinarity by bringing many fields of research together, eliminating silos and creating systematic cohesion and thinking. Convergence can also be demonstrated in an institutional context when universities and colleges build core facilities to better manage and maximize the shared use of their infrastructure, combine their strategic research priorities and research facilities to take on specific challenges and develop partnerships around the world. As well, convergence can be understood as the deepening collaboration between researchers and research organizations in academia, the private sector, government and non-governmental organizations that share an interest in addressing social, economic and environmental challenges, fostering innovation, and improving quality of life.
>
> (CFI, 2017; pp. 4–5)

As we enter the fourth industrial revolution that blurs the boundaries between the biological, physical, social, and digital realms (Floridi, 2014), digitization of both everyday life and of the agri-food system may be a key transformational lever for societal-scale solutions to grand challenges. The unprecedented innovation power, speed, and connectivity it brings may accelerate the engagement of actors operating at community, city, state, and global level on paths of integrated sustainable development, targeting more convergence between economic, environmental, health, and other social outcomes (Block, 2016; Dube et al., 2012, 2014c; Hammond & Dube, 2012). This chapter focusses on how digitization can accelerate convergence in science and innovation, examining how it can support design, operation, and adaptive learning for the system as a whole and for actors along supply chains and markets. We introduce convergence science and innovation (CI) as a framework that places the human beings themselves at the center of transformation needed on supply and demand sides of the social and economic divide that has structured development since the onset of the first industrial revolution (Dube et al., 2012, 2014b, 2018; Jha et al., 2016).

Beyond system-level transformation, changes in many everyday human behaviors and responsible action on the personal and professional sides of life are pressing prerequisite to solutions at scale (Blok et al., 2016). A person is often facing conflict more than convergence among his/her roles of consumer, producers, patients, and citizens (Dube et al., 2012, 2014a, 2014b). Yet, current approaches to science and innovation operate either target systems or individual- without keeping individuals nested in higher levels that define contexts (Hammond & Dube, 2012). In this chapter, we briefly review key tenets of CI. We then review how recent advances in data and digital technologies are making it possible through building capability to (1) develop deep understanding of food choices and behavior through behavioral analytics, (2) articulate a person-centered systems approach to link individual decision making to the multi-scale convergence of technical, social, and institutional innovation for societal-scale supply and demand solutions for healthy diets from sustainable food systems, and (3) offer a multi-stakeholder approach for digital support to agri-food innovation.

Convergence research and innovation (CI)

CI proposes to innovate the way we innovate by highlighting the fact that successful innovations with substantial economic and societal impacts tend to consist of a convergence of technological, business, social, and institutional innovations (see Figure 10.1). In the agri-food space, examples of technological innovations include innovations in agricultural inputs and methods, scientific or technical innovations in food processing and distribution, and even behavioral innovations (e.g., relating to nutritional intake, wellness, healthcare, and interactions with the environment). Adoption of technological innovations often requires the development of suitable business models to produce, deliver, and maintain the innovations; at times creating or disrupting entire value chains. Especially when innovations are disruptive, they tend to be accompanied by changes to social routines, networks, at times even beliefs and attitudes – that is, social innovations in the broad sense of the term. Vastly disruptive innovations, such as those related to genetically modified organisms or gene editing, also elicit institutional responses, both formal and informal, involving legislative, policy, and program changes, and potentially affecting economic frameworks and market designs.

Applying the CI perspective to innovation in the agri-food space and the societal challenges outlined above, CI aims to provide a framework to bridge or leapfrog the still prevailing social-economic divide structuring society (Dube et al., 2012, 2014b, 2018; Jha et al., 2016) while averting a potential digital divide (Bronson & Knezevic, 2019). CI targets not only the institutional transformation but also innovation at the level of technologies, communities, supply chains, markets, and other organizations and systems that form economy and society. In other words, CI combines technological, business, social, and institutional innovation to reach societal scale integrated outcomes through both

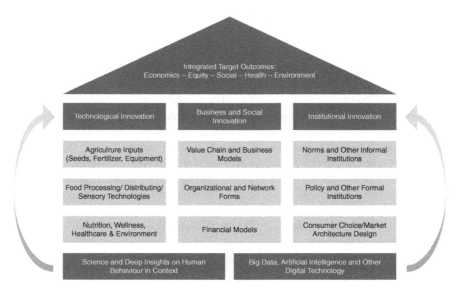

Figure 10.1 Convergent innovation framework. Adapted from Dube et al. (2014b).

individual and collective action of actors throughout society. This includes the person in his/her diverse role as consumer, citizen, producer, patients, or other as well as professions, organizations, and institutions (Figure 10.1).

The first key enabler of CI is the combination of recent advances scientific understanding of real-world human behavior, with the knowledge we have on intersectoral and multi-scale pathways creating real-world contexts. Combined, these provide insights for moving toward more adaptive action by the person themselves or all organizational and institutional actors involved. This departs from current state of affairs as most research and innovation efforts deployed to address grand challenges still either focus on systems design and policy, or at the individual level, primarily through education. This does not fully account for the diversity of roles and behaviors of a person in his/her everyday lives as consumers of private and public goods, services and programs, or as citizens, or else the producer roles played at community, organization, or institutional level. Beyond the roles occupied by a person through his/her life journey, science has also advanced on the different dimensions of real-world behavior (Doucerain & Fellows, 2012; Dube et al., 2010, 2016, 2018).

These cover both rational and non-rational components of human decision making: in addition to the well-known heuristic and biases in decision making under risk and uncertainty (object of original behavioral economics), the cue-induced/goal-driven decision making (current object of many marketing program), and planning/prospective judgment (current object of current grand challenge solution model), the other three tap into the socio-emotional components of behavior, namely sensory/embodied experience, emotion/stress responses, and interpersonal/social processes.

CI aims to achieve better impact in promoting adaptive behavior through more precise targeting of innovation and intervention to differentiated needs, expectations, and aspirations of individuals or population segments. In addition, changes in individual and collective belief systems and cultural values are happening, reducing the hegemony of short-term materialism that has fueled different types of externalities in modern society since the onset of the first industrial revolution is fading in favor of more balance with longer term ethical and societal concerns. The specification of most critical lever points for change in individual decision making, experience, and behavior becomes input for person-centered innovation and action throughout society. In sum, the underlying model characterizes biologically, psychologically, and culturally meaningful components of human decision making and behavior in diverse and dynamic real-world physical and digital contexts, and uses deep/machine learning and other computational models to link this scientific understanding of individual choice to the understanding of the societal pathways by which contexts exercise their short- and long-term impact (Dube et al., 2018; Hammond & Dube, 2012; Hammond et al., 2012; Hagen et al., 2020). CI builds upon these models to embed this knowledge into and enrich disciplinary and sectoral science, innovation, process, and practice at professional, organizational, institutional, or system in all domains that contribute to individual and collective health, wealth, and well-being. This brings a person-centric approach to organizations' innovation pipelines, marketing, and business strategies as well as to the systems and institution within which these are embedded.

Digitization is a key enabler for tackling this multi-scale complexity in a tractable and solution-oriented manner as big data, artificial intelligence, and other digital technologies have become ubiquitous. Beyond its anchor role in precision agricultural at the farm level, digitization throughout the whole agri-food systems, and broader society connects in unprecedented way people, organizations, and institutions as well as the data, models, and mindsets that support their decisions and actions. Embedded with value-sensitive innovation and action on the ground, digitization may accelerate the bringing down of disciplinary and sectoral silos for single and collective action targeting the convergence of economic, social, environmental, and health outcomes, as is required for grand "wicked" challenges. These silos to break down or leapfrog may be separating social sciences from their life, natural and engineering sciences counterparts, or separating societal (private, civil society, and public) and industrial (e.g., farm, food, transportation, energy, bio tech/pharma, digital) sectors, or else these separating science from its transfer to societal and industrial sectors for short and long-term contribution to economy and society as per the traditional linear models of technical innovation. Digitization is clearly a key enabler of accelerated transformation to responsible eating for the consumer and professionalism in the agri-food sector. It has to be however appropriately designed and deployed in a value-sensitive manner and embedded with real-life actions in a person-centered manner that captures diversity and dynamics in motives, contexts, and conditions (Blok, 2018; Dube et al., 2014b).

To account for the diversity of actors involved and the scale at which either or both behavioral change and ecosystem transformation are needed for integrative sustainable development, CI proposes the creation of modular platforms and project portfolios. Modularity (Sanchez & Mahoney, 1996; Schilling, 2000) enables bridging the social and economic divide in a manner that supports individual and collective value creation. Modular projects cover the full spectrum of private, pre-competitive, and public value creation targeting scopable and achievable solutions, regardless of their origin on one or the other side of the social-economic divide (Jha et al., 2016). Altogether providing societal-scale solution while pursuing single and collaborative goals (Dube et al., 2012, 2014a, 2018), scientists and action partners from diverse disciplines and sectors at local, national, and global communities are assembled around targeted and achievable goals at the convergence of human, social, and economic development with innovation and intervention design being informed by deep understanding of individuals and contexts.

Behavioral analytics

The ability to achieve individual and collective goals for integrative sustainable development hinges on the normative and adaptive quality of human behavior, be it at professional, organizational, institutional, system, or policy context (Cajaiba-Santana, 2014). Changes in everyday human behaviors in their personal lives are also a pressing prerequisite for societal scale solutions (Pol & Ville, 2009; van der Have & Rubalcaba, 2016).

Large-scale data on human behavior in food, diet, lifestyle, and health domains are becoming more available, opening new perspectives on bridging biological, social, and food environments. Promising methods applying consumer insights, behavioral economic nudges, and other theory-based behavioral assessments and change strategies are increasingly informing interventions targeting health-promoting dietary behavior (Broers et al., 2017). Consumer insights, for instance, combined with consumer journey mapping tools, provide information on the various drivers of eating behavior across individuals, populations, situations, time, as well as lifestyle and cultural values (Martin & Morich, 2011). Consumer insights also provide input into how nutrition and health are positioned in relation to other motives such as taste, fun, convenience, price, or others. In commercial contexts, they identify points of value creation along the full experience of shopping, purchasing, preparing, consuming, or disposing of food products (Jeltema et al., 2015; Sukhdial & Boush, 2004). Loyalty and relationship management programs of manufacturers and/or retailers encourage the repetition of such cycles and provide opportunity for assessment of long-term patterns and outcomes.

New digital methods of behavior analytics integrate large-scale data on the diverse, dynamic, and oftentimes conflicting drivers of individual and household

Convergence research and the digital backbone 117

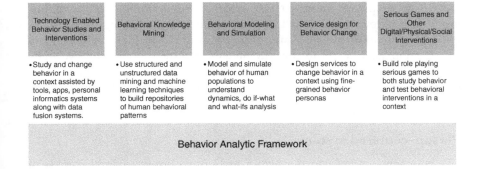

Figure 10.2 Behavior analytic framework.

dietary behaviors to develop deep/machine learning and other AI methods to better identify underlying behavioral patterns that were previously undetectable using traditional statistical analysis, along with their relationship to biological, social, and food systems (Dube et al., 2018). In addition, the integration of consumer insights and behavioral economics may help to design and deploy interventions targeting lifelong nutrition, health, and wellness in a manner that is also economically, culturally, environmentally sustainable (Arora et al., 2014; Dube et al., 2014a). The emerging rich pool of real-time behavior data sets, enabled by the digitization of everyday life, economy, and society, are now available at an unprecedented level of contextual, spatial, and temporal granularity (Groves et al., 2016).

The behavioral analytics framework (Figure 10.2) can help explain how behavioral insights can be derived from large-scale data using predictive and other models to inform dietary choices and, in some cases, monitor their long-term outcome. This can be in terms of behavioral components along the consumer journey or in terms of influences at different temporal or geographical scales that define choice contexts (Dube et al., 2014b). In the food, nutrition, and diet domains, the proliferation of apps that support consumers range from shopping aids, through consumption records, to goal setting and monitoring, oftentimes combined with wearable technologies. The richness of data then allows for the development of health nudges or other interventions, such as using social or goal-based reference points to motivate individuals to adopt healthier lifestyles (Uetake & Yang, 2018, 2019). For example, recent consumer research shows that users of FitBit exercise more if their network of friends also exercises more, compounding the positive effect of the digital support (Brown, 2010). These studies provide a first glimpse of the possibility to leverage big data in developing health-related nudges. The approach aims not only to characterize differentiated behavioral patterns and/or predict outcomes, but also to trace underlying mechanisms that can form the basis of interventions (Figure 10.2).

Finally, behavioral analytics, AI, and other digital technologies can inform the design of theory-informed and evidence-based health/food/nutrition-promoting innovation and/or interventions, be these of a digital, social, or physical nature. For instance, large-scale data can aid in the development of gamification/goal design for healthy nudges, as shown in an extensive study about millions of Lose It mobile users, another popular app supporting healthy diet and lifestyle with a primary goal of weight management (Pappa et al., 2017).

Person-centered systems

The emergence of behavioral analytics applications points to the possibility of next generation decision support tools. These tools would support not only individuals but also provide person-centered design, deployment, and monitoring insights enabling all actors throughout society to define the contexts in which a person's food choices and dietary behaviors are performed. Beyond digitalization at the individual level, big data, AI, and other technologies have become ubiquitous in innovation, strategy, and practice at multi-level analysis across all societal levels. Can the scientific, commercial, and societal benefits from agri-food science and technology be scaled up if bridged with a research program that brings in behavioral economics and consumer choice/market response modeling and is supported by big data and predictive analytics models that specifically target convergence between economic, social, health, and environmental outcomes? This would require equal scientific rigor and investment in the study of demand drivers for healthy diets, and the translation of these drivers into successful sustainable food innovation and commercialization. The brain-to-society (BtS) choice model (Dube et al., 2008, 2010; Hammond & Dube, 2012) is insightful in this regard.

Neuroscience has typically considered individual components of behavior one at a time, divorced from the complexity of the social and physical environment, a key component to understanding real-world behaviors. The BtS model bridges the frontiers of knowledge in the behavioral, social, and economic sciences, to those of neuroscience, data, computer, and complexity sciences, with the aim to develop multi-scale and solution-oriented tools that translate science into practice, and vice-versa (Dalle Molle, 2017; Dube et al., 2008, 2010; Hammond & Dube, 2012). The model is anchored in robust theoretical understanding and formal models of underlying genes and brain systems guiding individual choice and behavior, including gene–environment (G*E) interaction and precision health concepts. Together with complexity and computational sciences this model can help understand causality within both individuals and in terms of the combined contribution of the complex and dynamic web of biological, social, and system levels factors impacting a person's dietary behavior. As illustrated in Figure 10.3 (Hammond & Dube, 2012), individuals are seen as being nested within organizations and systems that define contexts in which they operate.

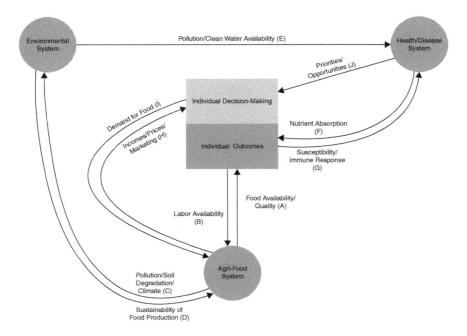

Figure 10.3 A systems framework for food and nutrition security.

This brings a person-centric approach to organizations' innovation pipelines, marketing and business strategies, as well as to the systems and policy within which these are embedded. For instance, combined with biological platforms ranging from microbiome, to genomics, to digital taste and other facets of the food products, and eating experiences, behavioral insights can translate into differentiated product features and characteristics promoting adaptive behavior in targeted individuals or population segments (Portella et al., 2019). Behavioral and context insights generated are then utilized by decision-makers from businesses, public agencies, and community organizations and institutions to adopt a person-centered approach to their individual and collective actions. Increasingly, the use of behavioral data and insights is also governed by privacy and data ownership legislation.

One particular type of analytics tools in bridging what is in consumers' minds to support adaptive behavior through convergence innovation systems design is natural language processing, and deep/machine learning model of unstructured text and visual data from different sources (Dube et al., 2018; Ferro et al., 2018), including data collected from social media, as well as data from companies' websites, newsfeeds, and consumer cards.

A second type of analytics tools are digital synthetic ecosystems (SE) that position individuals within the complex and dynamic contexts affecting individuals' behavior (Cajka et al., 2010), as illustrated in Figure 10.4 (adapted from

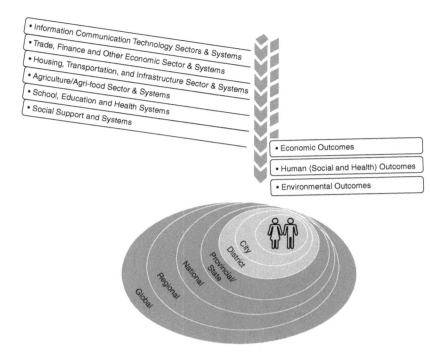

Figure 10.4 Person-centered ecosystem.

Dube et al., 2014a). Researchers use SE to visualize scenarios which cannot be carried out in real populations or for which adequate historical data on natural experiments are not available. For instance, we are developing for the City of Montreal a SynthEco platform. This virtual platform permits the development of SE – statistically representative synthetic populations and environments and the simulation of the impact of different intervention prototypes over time. This is achieved by combining various disparate data collection efforts, such as census, cohorts, clinical studies, and diverse surveillance data, into a population-level, geographically explicit representation to equip government, private, and academic research for population-level planning. Traditionally, SEs have served as the basis for agent-based simulation in infectious disease and public health modeling, as well as transport modeling (Bazzan et al., 2015; Monteiro et al., 2014; Wang & Tang, 2004; Yu et al., 2014).

Multi-stakeholder decision support to agri-food innovation

The person-centered approach outlined above emphasizes that food choices and behaviors of individuals are intrinsically connected to multi-level, multi-stakeholder agri-food ecosystems. In turn, this implies that innovation and commercialization in agri-food ecosystems is potentially affected by all levels of the model, which raises particular challenges if agri-food innovation aims to promote behavior centered on adaptive diets from sustainable systems. Digitalization, shared standards and protocols combined into multi-stakeholder digital architectures are set to overcome such problems with an integrating and transdisciplinary collaboration to stimulate open science, open access, and open to the world. Sharing data, facilities, tools, and expertise enables breakthroughs in research, thereby supporting policy and innovation in the field of sustainable food, nutrition, and health (Musker et al., 2018) and potentially shaping public debate on food.

Multi-stakeholder digital platforms and research infrastructures to support agri-food innovation are currently emerging around the world. Their drivers are the trends outlined above: (1) societal challenges related to food production and consumption, such as sustainability and security issues, climate change, and the increase in lifestyle diseases (linked to preventive health); (2) rapidly developing digital technologies and behavioral analytics; (3) the emergence of person-centered approaches, and they are further enabled by (4) emerging data hub expertise, which enhances know-how on how to combine data from different sources and formats; and (5) political strategy that promotes collaboration in research infrastructure settings.

For example, the SmartAgriHubs in development across the European Union (EU) embrace a demand-driven growth methodology for multi-stakeholder digital support to agri-food innovations. Here, end-users from the agri-food sector are driving the growth of the Digital Innovation Hubs (DIHs) network through innovation experiments. The process builds on an existing network and ecosystem and requires that basic hardware and software applications be put into place. There is also a concurrent need to ensure that data can be acquired and stored with increased security, enacting standards for these new data types and to provide relevant analytics capabilities (Musker & Schaap, 2018).

Figure 10.5 demonstrates the five concepts of the SmartAgriHubs approach: (1) Competence Centers form the cornerstone for DIHs, where expertise, infrastructures, etc. are available; (2) DIHs, through which the competences are matched with demands, ideas, funding, etc. and orchestrated and supported by concrete services to translate this interaction; (3) Innovation Experiments, in which ideas, concepts, prototypes, etc. are further developed, tested, and finally introduced into the market; (4) Innovation Services Maturity Model monitoring, assessing, and helping the DIHs' innovation services to reach their desired level; (5) an Innovation Portal as a searchable register for knowledge exchange, brokerage, etc. To manage the network, a layer of Regional Clusters are envisioned, to be coordinated by the central project management.

Platforms like the SmartAgriHubs afford the opportunity to actively engage the multiple stakeholders of the agri-food ecosystems meaningfully in the innovation process (Schillo & Robinson, 2017). Implemented responsibly and effectively, they may contribute to transformations that bridge the many divides that remain in the agri-food system; for example, those between consumers and

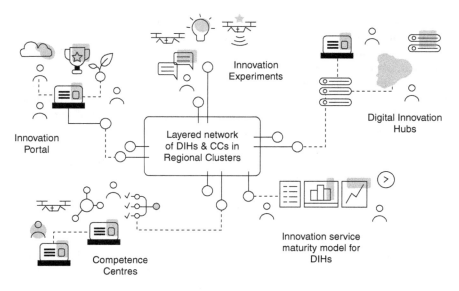

Figure 10.5 Layered network of digital innovation hubs and competence centers.

producers, science and its update, and in a larger sense between individual and collective value creation and between economic, ecological, and social processes, outcomes, and responsibilities. Multi-stakeholder platforms may also provide an opportunity to articulate and visualize differences in agri-food systems around the world, and engage in dialogue and trade to bridge divides between the West and the Rest, a divide that persists as countries around the world are now trying to strike a better balance between traditional and modern economy and society (Dube et al., 2012).

Conclusion

Convergence research and innovation may sketch a powerful distributed alternative to the still prevailing vertically integrated model of agri-food systems. Digitalization linking biological, social, and commercial dimensions of agri-food systems offers the opportunity to develop frameworks and platforms that transcend the perspectives of individual actors, that integrate ethical, social, environmental, and commercial concerns, and that can legitimately aspire to do so at a societal level and scale. This chapter has outlined several frameworks and platforms currently being applied and developed further, and which, over time, may help identify the leverage points that will allow the current societal-scale transformations of the fourth industrial revolution alleviate negative externalities while preserving progress sparked by first industrial revolution mode of development around the world. Realizing the SDGs demands not only a technological "fix" for our problems, but also an encompassing societal change in human choices, actions, interactions, and in the organization of society, including its policy.

References

Ajay, S., & Boush, D. M. (2004)."Eating Guilt: Measurement and Relevance to Consumer Behavior" in B. E. Kahn & M. F. Luce (Eds.), *Advances in Consumer Research Volume 31*. Valdosta, GA: Association for Consumer Research, pp. 575–576.
Arora, N. K., Pillai, R., Dasgupta, R., & Garg, P. R. (2014). Whole-of-society monitoring framework for sugar, salt, and fat consumption and noncommunicable diseases in India. *Annals of the New York Academy of Sciences*, 1331(1), 157–173.
Bazzan, A. L., Heinen, M., & do Amarante, M. D. B. (2015). *ITSUMO: An Agent-Based Simulator for Intelligent Transportation Systems*. In *Advances in Artificial Transportation Systems and Simulation* (pp. 1–20). San Diego, California: Elsevier/Academic Press.
Blok, V., Gremmen, B., & Wesselink, R. (2016). Dealing with the wicked problem of sustainability: the role of individual virtuous competence. *Business and Professional Ethics Journal*, 34(3), 297–327.
Blok, V. (2018). Beyond technocratic management in the food chain – towards a new responsible professionalism in the Anthropocene. In Springer, S., & Grimm H. (Eds.), *Professionals in food chains* (pp. 411–419). Wageningen: Wageningen Academic Publishers.
Broers, V. J., De Breucker, C., Van den Broucke, S., & Luminet, O. (2017). A systematic review and meta-analysis of the effectiveness of nudging to increase fruit and vegetable choice. *The European Journal of Public Health*, 27(5), 912–920.
Bronson, K., & Knezevic, I. (2019). The digital divide and how it matters for Canadian food system equity. *Canadian Journal of Communication*, 44(2), 63–68.
Brown, S. J. (2010). *Social network for affecting personal behavior*. Google Patents, Application 12/800,343, filed September 16, 2010.
Cajaiba-Santana, G. (2014). Social innovation: moving the field forward. A conceptual framework. *Technological Forecasting and Social Change*, 82, 42–51.
Cajka, J. C., Cooley, P. C., & Wheaton, W. D. (2010). Attribute assignment to a synthetic population in support of agent-based disease modeling. *Methods Report* (RTI Press), 19(1009), 1.
Canada Foundation for Innovation, Corporate plan 2019–20. Retrieved from www.innovation.ca/sites/default/files/pdf/2019-20_cfi_corporate_plan.pdf
Dalle Molle, R., H. et al. (2017). Gene and environment interaction: is the differential susceptibility hypothesis relevant for obesity? *Neuroscience & Biobehavioral Reviews*, 73, 326–339.
Doucerain, M., & Fellows, L. K. (2012). Eating right: linking food-related decision-making concepts from neuroscience, psychology, and education. *Mind, Brain, and Education*, 6(4), 206–219.
Dubé, L., Bechara, A., Böckenholt, U., Ansari, A., Dagher, A., Daniel, M., ... & Huettel, S. (2008). Towards a brain-to-society systems model of individual choice. *Marketing Letters*, 19(3–4), 323–336.
Dubé, L. (2010). *Obesity prevention: the role of brain and society on individual behavior*. 1st ed, Amsterdam: Academic Press.
Dubé, L., Pingali, P., & Webb, P. (2012). Paths of convergence for agriculture, health, and wealth. *Proceedings of the National Academy of Sciences*, 109(31), 12294–12301.

Dubé, L., Addy, N. A., Blouin, C., & Drager, N. (2014a). From policy coherence to 21st century convergence: a whole-of-society paradigm of human and economic development. *Annals of the New York Academy of Sciences*, 1331(1), 201–215.

Dubé, L., Jha, S., Faber, A., Struben, J., London, T., Mohapatra, A., ... & McDermott, J. (2014b). Convergent innovation for sustainable economic growth and affordable universal health care: innovating the way we innovate. *Annals of the New York Academy of Sciences*, 1331(1), 119–141.

Dubé, L., Labban, A., Moubarac, J. C., Heslop, G., Ma, Y., & Paquet, C. (2014c). A nutrition/health mindset on commercial Big Data and drivers of food demand in modern and traditional systems. *Annals of the New York Academy of Sciences*, 1331(1), 278–295.

Dubé, L., Du, P., McRae, C., Sharma, N., Jayaraman, S., & Nie, J.Y. (2018). Convergent innovation in food through big data and artificial intelligence for societal-scale inclusive growth. *Technology Innovation Management Review*, 8(2), 13–29.

Ferro, N., Fuhr, N., Nie, J.Y., Grefenstette, G., Konstan, J. A., Castells, P., Daly, E. M., ... Zobel, J. (Eds.) (2018). From evaluating to forecasting performance: how to turn information retrieval, natural language processing and recommender systems into predictive sciences. *Dagstuhl Manifestos*, 7(1), 96–139.

Floridi, L. (2014). *The fourth revolution: How the infosphere is reshaping human reality.* Oxford University Press: Oxford.

Groves, P., Kayyali, B., Knott, D., & Kuiken, S. V. (2016). *The big data revolution in healthcare: accelerating value and innovation.* McKinsey.

Hagen, L., Uetake, K., Yang, N., Bollinger, B., Chaney, A., Dzyabura, D., Etkin, J., Goldfarb, A., Liu, L., Sudhir, K., Wang, Y., Wright, J., & Zhu, Y. (2020). How Can Machine Learning Aid Behavioral Marketing Research? *Marketing Letters*, forthcoming.

Hammond, R. A., & Dubé, L. (2012). A systems science perspective and transdisciplinary models for food and nutrition security. *Proceedings of the National Academy of Sciences*, 109(31), 12356–12363.

Hammond, R. A., Ornstein, J. T., Fellows, L. K., Dubé, L., Levitan, R., & Dagher, A. (2012). A model of food reward learning with dynamic reward exposure. *Frontiers in Computational Neuroscience*, 6, 82.

James, P.T., Rigby, N., Leach, R., & International Obesity Task Force. (2004). The obesity epidemic, metabolic syndrome and future prevention strategies. *European Journal of Cardiovascular Prevention & Rehabilitation*, 11(1), 3–8.

Jeltema, M., Beckley, J., & Vahalik, J. (2015). Model for understanding consumer textural food choice. *Food Science & Nutrition*, 3(3), 202–212.

Jha, S. K., Gold, R., & Dube, L. (2016). Convergent innovation platform to address complex social problems: a tiered governance model. *Academy of Management Proceedings*, 2016(1), 15150.

Kochanek, K. D., Murphy, S., Xu, J., & Arias, E. (2017). *Mortality in the United States, 2016.* NCHS Data Brief (293), 1–8.

Martin, N., & Morich, K. (2011). Unconscious mental processes in consumer choice: toward a new model of consumer behavior. *Journal of Brand Management*, 18(7), 483–505.

Monteiro, N., Rossetti, R., Campos, P., & Kokkinogenis, Z. (2014). A framework for a multimodal transportation network: an agent-based model approach. *Transportation Research Procedia*, 4, 213–227.

Musker, R., & Schaap, B. (2018). Global Open Data in Agriculture and Nutrition (GODAN) initiative partner network analysis. *F1000Research*, 7.

Musker, R., Tumeo, J., Schaap, B., & Parr, M. (2018). GODAN's impact 2014–2018 – improving agriculture, food and nutrition with open data. *F1000Research*, 7.

Must, A., Spadano, J., Coakley, E. H., Field, A. E., Colditz, G., & Dietz, W. H. (1999). The disease burden associated with overweight and obesity. *Jama*, 282(16), 1523–1529.

Pappa, G. L., Cunha, T. O., Bicalho, P. V., Ribeiro, A., Silva, A. P. C., Meira Jr, W., & Beleigoli, A. M. R. (2017). Factors associated with weight change in online weight management communities: a case study in the LoseIt Reddit community. *Journal of Medical Internet Research*, 19(1), e17.

Pol, E., & Ville, S. (2009). Social innovation: buzz word or enduring term? *The Journal of Socio-Economics*, 38(6), 878–885.

Portella, A. K., Paquet, C., Bischoff, A. R., Dalle Molle, R., Faber, A., Moore, S., ... & Dube, L. (2019). Multi-behavioral obesogenic phenotypes among school-aged boys and girls along the birth weight continuum. *PloS One*, 14(2), e0212290.

Rodd, C., & Sharma, A. K. (2016). Recent trends in the prevalence of overweight and obesity among Canadian children. *CMAJ*, 188(13), E313–E320.

Sanchez, R., & Mahoney, J. T. (1996). Modularity, flexibility, and knowledge management in product and organization design. *Strategic Management Journal*, 17(S2), 63–76.

Schilling, M. A. (2000). Toward a general modular systems theory and its application to interfirm product modularity. *Academy of Management Review*, 25(2), 312–334.

Schillo, R. S., & Robinson, R. M. (2017). Inclusive innovation in developed countries: the who, what, why, and how. *Technology Innovation Management Review*, 7(7), 34–46.

Subramanian, S. V., Kawachi, I., & Smith, G. D. (2007). Income inequality and the double burden of under-and overnutrition in India. *Journal of Epidemiology & Community Health*, 61(9), 802–809.

Sukhdial, A., & Boush, D. M. (2004). Eating guilt: measurement and relevance to consumer behavior. ACR North American Advances.

The United Nations (2017). *The Sustainable Development Goals Report 2017*. Department of Economic and Social Affairs of the United Nations Secretariat, New York.

Uetake, K., & Yang, N. (2018). Harnessing the Small Victories: Goal Design Strategies for a Mobile Calorie and Weight Loss Tracking Application. *Working paper*.

Uetake, K., & Yang, N. (2019). Inspiration from the "Biggest Loser": social interactions in a weight loss program. *Marketing Science*, 39(3), 487–499.

van der Have, R. P., & Rubalcaba, L. (2016). Social innovation research: an emerging area of innovation studies? *Research Policy*, 45(9), 1923–1935.

Wang, F. Y., & Tang, S. (2004). Artificial societies for integrated and sustainable development of metropolitan systems. *IEEE Intelligent Systems*, 19(4), 82–87.

Weitz, N., Nilsson, M., & Davis, M. (2014). A nexus approach to the post-2015 agenda: formulating integrated water, energy, and food SDGs. *SAIS Review of International Affairs*, 34(2), 37–50.

Yu, Y., El Kamel, A., Gong, G., & Li, F. (2014). Multi-agent based modeling and simulation of microscopic traffic in virtual reality system. *Simulation Modelling Practice and Theory*, 45, 62–79.

Conclusion
Where next?

Federica Brunetta and Maria Carmela Annosi

Digitalization in agri-food: where next?

The purpose of this book was to tackle how firms operating in the agri-food sector organize their handling of the challenges and opportunities offered by digital transformation. We have focused on providing contributions that included different theoretical perspectives and varied empirical backgrounds to describe, analyze, and interpret different phenomena, processes, and practices related to digital transformation. Since we have focused on digital transformation as "the use of new digital technologies (social media, mobile, analytics, or embedded devices) to enable major business improvements, such as enhancing customer experience, streamlining operations, or creating new business models" (Fitzgerald et al., 2014: 2), we have provided both theoretical and empirical perspectives to enrich this understanding, provided that an Agriculture 4.0 "era" has started and is quickly developing.

As one of our initial statements, following this examination of theory and evidence, we still argue that even in agri-food, "digital transformation is fundamentally not about technology, but about strategy" (Rogers 2016: 308): digital transformation has challenged incumbent firms and welcomed a plethora of new players with different roles. We have therefore addressed how managers have analyzed, interpreted, and eventually embraced this change, reformulating strategies and business models to accommodate new technologies. We have also focused on the increased attention to sustainability – both economic and environmental – that is putting more pressure on players to change.

The book has introduced and addressed critical issues related to (i) the challenges, opportunities, and problems created by digital transformation for agri-food firms; (ii) the impact of digital transformation on organizational capabilities, competencies, and routines; (iii) new business models; (iv) governance; and (v) economic and environmental sustainability. We have re-examined traditional organizational and strategic perspectives in light of the transformations that have led this sector into a renewed, information-rich, and networked industry, with players leveraging technology while countering its potential downsides.

In this final chapter, we do not aim to summarize elements that have been theorized throughout the book, or briefly recap the main findings, but rather we want to focus on the suggestion for potential avenues for research given the need to revise and develop new insights and theories on how organizations exist and function in this business context.

Avenues for research

Despite the significant benefits and the efforts of many institutions to pave the way for Smart Agriculture, there are still many organizations coping with challenges in adopting digitalization; this was a recurring topic throughout the book. While management scholars have significantly analyzed factors influencing its adoption, diffusion, and use, the specific context of agri-food still appears to be overlooked.

Generally, studies about adoption and diffusion address two different levels: studies about adoption emphasize the factors that involve if and how a particular organization, either in time or extent, will approach a new technology. On the other hand, diffusion is often interpreted as aggregate adoption by a potential market (e.g., the percentage of the farming population, or the total land, adopting an innovation). The pace and likelihood of both adoption and diffusion are traditionally affected by dynamic and uncertain processes that result in disruption to prices, learning, and use.

Nonetheless, this industry presents specific characteristics, and a significant number of factors appear to influence the decision-making process on the uptake – the adoption and adaptation – of agricultural technologies (Sunding and Zilberman, 2001). For instance, the agri-food industry's traditional competitive mechanism is moderated by a relevant number of institutional constraints and policies that might significantly influence strategic behavior. Policy and institutional support may be especially important in the area of technological change and adoption, through subsidies, support, infrastructure, and services. Of course, organizations and institutions are interdependent in some ways: institutional constraints may affect the patterns of adopting new technologies, but at the same time, the introduction of new technologies could affect the institutional structure and operation of agricultural industries.

Since technological innovation and institutional change are tightly interwoven, they both have profound effects on the evolution of the agri-food industry (Sunding and Zilberman, 2001). Agricultural economists clearly highlight that innovations do not occur randomly, but rather that incentives and government policies affect the nature and the rate of innovation and adoption (Sunding and Zilberman, 2001). Thus, an important point for future research is to further analyze institutional solutions and deepen the understanding of institutional innovations and their interdependence with digital transformation. This is especially relevant in terms of analyzing governance issues for platforms and ecosystems.

Networks and collaborations have also proven to be very important in this context. Through platforms, the value chain has become integrated and networking has been thoroughly examined as an organizational mechanism for coordinating and pursuing innovation adoption and diffusion processes. Nonetheless, these studies are still limited for the agri-food industry and are mostly focused on structural characteristics (Lambrecht at al., 2018). A better understanding of the role of each actor, the strength of their ties, and the innovation outcome could help shed new light on the relationship between networks and innovation in agri-food.

A fundamental issue that has emerged from our analysis is the constant tension between economic and environmental sustainability. We acknowledged the complexity of the industry and the need for innovation to ensure both objectives. In order to do so, agri-food systems must shift toward new paradigms, moving away from traditional business models and embracing digitalization. In this light, many alternative forms of agriculture (e.g., cooperatives, circular economy, platforms, social farming, urban farming) have emerged and crossed existing boundaries, establishing diverse institutional logics. Agricultural Innovation Ecosystems (AIE) have responded to the need for interaction between diverse players, but the analysis of this tension undoubtedly requires additional understanding (Walrave et al., 2018).

Following the three highlighted points (institutions, networks, and collaboration), we argued that scholars should better focus on why and how firms adopt new practices, in addition to new technologies. Preliminary work has been done to derive organizational design recommendations regarding new forms of organizing for digitalization, pointing at the need to increase organizations' agility in detecting environmental change. However, past studies have indicated some issues in the implementation of agility principles (see Annosi and Brunetta, 2018; Annosi et al., 2017; Annosi et al., 2018). This calls for more research studies on analyzing how small and medium firm in agri-food sector can organize their business and how institutional environment can facilitate the development of firm's digital capabilities.

Scholarly contributions have offered important insights into diffusion processes among groups of organizations, mostly following two different approaches (Strang and Macy, 2001): the first one is based on the rational-actor model, in which adoption is recognized as driven by efficiency gains and productivity enhancement, ultimately leading to superior performance (e.g., Katz and Shapiro, 1987). The second one is based on a social and institutional perspective, with motivation driven by the quest for legitimacy in the eyes of peers, stakeholders, partners, and customers, given an individual's place within larger groups (DiMaggio and Powell, 1983). Some contributions (e.g., Tolbert and Zucker, 1983) have even highlighted a two-stage model where the rational explanation of early adopters interweaves with the quest for legitimacy of later adopters.

Nonetheless, while most studies have focused on the motivations driven by the characteristics of adopters or follow-the-trend behaviors in adopting innovation, a lack of research exists in exploring adopters' motivations. Thus, further research should be carried out by reconciling economic and social explanations of diffusion by looking closer at the decisions and motivations behind adoption, even in contexts where institutions drive processes. We feel that the agri-food industry, given its characteristics, would be an ideal field for analyzing and deepening scholarly understanding of these issues. This also leaves open an interesting issue in the relationship between institutions, interactions, and diffusion, that is, the pace and extent of implementing practices. Even if organizations are following needs for efficiency or legitimacy when adopting new technologies, they will still need to implement new practices into existing operations and political agendas. Several authors are highlighting the need to view diffusing practices as dynamic (Rogers, 1995; Strang and Soule, 1998) and, in this light, to analyze the details of implementation (e.g., Fiss and Zajac, 2006; Westphal and Zajac, 2001), especially since authors (Westphal et al., 1997) have been arguing that efficiency requires the implementation of customized practices, while legitimacy concerns traditionally-driven isomorphic behaviors when it comes to implementing practices.

To summarize, we argue that the agri-food industry presents dynamics that are interesting for scholars, given its peculiar characteristics in terms of institutional, societal, and competitive dynamics as well as considering the deep transformation that digitalization is causing. We hope that our readers will be able to see the connections between each chapter and best understand the potential avenues for research. Of course, our main aim was to offer an overview of the main issues and critical points, identifying overlaps among different approaches, literature, settings, and theories.

References

Annosi, M.C., Foss, N.J., Brunetta, F., & Magnusson, M. (2017). The interaction of control systems and stakeholder networks in shaping the identities of self-managed teams. *Organization Studies*, 38(5), 619–645. ISSN: 0170-8406.

Annosi, M.C., & Brunetta, F. (2018). Resolving the dilemma between team autonomy and control in a post-bureaucratic era. Evidences from a telco multinational company. *Organizational Dynamics*, 47(4), 250–258. ISSN: 0090-2616.

Annosi, M.C., Martini, A., Brunetta, F., & Marchegiani, L. (2018). Learning in an agile setting: a multilevel study on the evolution of organizational routines. *Journal of Business Research.* 110: 554–556. ISSN: 0148-2963.

DiMaggio, P.J., & Powell, W.W. (1983). The iron cage revisited: institutional isomorphism and collective rationality in organizational fields. *American Sociological Review*, 147–160.

Fiss, P.C., & Zajac, E.J. (2006). The symbolic management of strategic change: sensegiving via framing and decoupling. *Academy of Management Journal*, 49(6), 1173–1193.

Fitzgerald, M., Kruschwitz, N., Bonnet, D., & Welch, M. (2014). Embracing digital technology: A new strategic imperative. *MIT Sloan Management Review*, 55(2), 1.

Katz, M.L., & Shapiro, C. (1987). R and D rivalry with licensing or imitation. *The American Economic Review*, 402–420.

Lambrecht, E., Crivits, M., Lauwers, L., & Gellynck, X. (2018). Identifying key network characteristics for agricultural innovation: a multisectoral case study approach. *Outlook on Agriculture*, 47(1), 19–26.

Rogers, E.M. (1995). *Diffusion of Innovations (4th ed.)*. New York: The Free Press.

Rogers, D.L. (2016). *The Digital Transformation Playbook: Rethink Your Business for the Digital Age*. New York: Columbia University Press.

Strang, D., & Macy, M.W. (2001). In search of excellence: fads, success stories, and adaptive emulation. *American Journal of Sociology*, 107(1), 147–182.

Strang, D., & Soule, S.A. (1998). Diffusion in organizations and social movements: From hybrid corn to poison pills. *Annual Review of Sociology*, 24(1), 265–290.

Sunding, D., & Zilberman, D. (2001). The agricultural innovation process: research and technology adoption in a changing agricultural sector. *Handbook of Agricultural Economics*, 1, 207–261.

Teece, D.J. (2010). Business models, business strategy and innovation. *Long Range Planning*, 43(2–3), 172–194.

Tolbert, P.S., & Zucker, L.G. (1983). Institutional sources of change in the formal structure of organizations: the diffusion of civil service reform, 1880–1935. *Administrative Science Quarterly*, 28(1), 22–39.

Walrave, B., Talmar, M., Podoynitsyna, K.S., Romme, A.G.L., & Verbong, G.P. (2018). A multi-level perspective on innovation ecosystems for path-breaking innovation. *Technological Forecasting and Social Change*, 136, 103–113.

Westphal, J.D., & Zajac, E.J. (2001). Decoupling policy from practice: the case of stock repurchase programs. *Administrative Science Quarterly*, 46(2), 202–228.

Westphal, J.D., Gulati, R., & Shortell, S.M. (1997). Customization or conformity? An institutional and network perspective on the content and consequences of TQM adoption. *Administrative Science Quarterly*, 42(2), 366–394.

Index

Academic Journal Guide 2018 12
advisory boards 64–65
Agricultural Innovation Ecosystems (AIE) 128
"Agriculture 4.0" 26
"Agrifood 4.0" 25
Agri Food Chain Coalition (AFCC) 78
AI *see* artificial intelligence (AI)
Airbnb 68
Amazon 108
artificial intelligence (AI): chatbots 104; crop and soil monitoring 103; data analysis and 102; disruptiveness of 101–2; Elaisian and 109–10; evolution of 6; generally 3, 62, 101–2, 104–5; historical background 101; main applications of 102–3; Precision Agriculture Technologies (PAT) and 103–4; predictive analytics 103; retail and food delivery and 104; robots 27, 103; supply chains and 102–4; support of agri-food through 103–4; sustainability and 6
automation 3

Bayer/Monsanto 107–8
behavioral analytics: in agri-food context 116–18; convergence research and innovation (CI) and 116–18; evolution of 6; framework 117, *117*; sustainability and 6
behavior, managerial 42–3
beliefs, managerial 39–41
"best practice" business processes 2
Big Data Analytics (BDA) 15–17, 26
Big Data (BD) 15–17, 62, 73
Biotechnology Industry Research Assistance Council of India 94
blockchain technology 3, 62, 73
BM *see* business models (BM)

brain-to-society (BtS) choice model 118, *119*
business ecosystems: actors participating in 63; adaptation in 70; advisory boards 64–5; in agri-food context 77–8; business models (BM) in 68–9; collaboration in 74–5, 78, 83, 86–8; communication issues in 70–1; contractual governance 80; cooperation in 78–9; coopetition in 78–9; coordination in 64; defined 75; designers 77; disruptiveness of technology in 68; dynamics of 4–6, 82–8; ecosystem-as-affiliation 78; ecosystem-as-structure 78; emerging questions 5; ethics and 63; evolution of 69–72, 76; external factors in 85–6; former projects 73; future trends 73; generally 88; generic schema of *79*; governance of 64–6, 75–6, 79–82; ideators 77; information transfer in 63; intermediaries 77; internal contingencies in 86–7; keystone players 76; layers of 77; lead organization governance 82, 85; legal and regulatory effects 63; lifecycle 84; major drivers for joining 63–4; Network Administrative Organization (NAO) governance 81, 85, 87; networks and 74–6; niche players 76; pervasiveness of technology in 68; phases of 75; physical dominators 76; platforms 76; principles governing interactions 65–6; relational governance 80; research gaps 83; shared governance 81–2, 85; stages of 84; sustainability of 72–3; technology trials 67; tensions in 66; transactions among actors in 69; trust in 81; value dominators 76

132 *Index*

business impact of digitalization: cognitive capacity, effect of 28, 31; discussion 28, 31–2; education, effect of 28, 31; finance, effect of 31; generally 25–6, 32; information transfer and 28; infrastructure, effect of 31; literature review 27–8, **29–30**
business models (BM): in business ecosystems 68–9; digitalization and 11–12; generally 1–2

Canada Fund for Innovation (CFI) 112
chatbots 104
climate change 109
Climate Corporation 107
Cloud Computing 26
cognition, managerial 41–2
collaboration: in business ecosystems 74–5, 78, 83, 86–8; case studies 51–2; digitalization and 18; future research 128; multi-stakeholder decision support and 121
Competence Centers 121, *122*
competition analysis 108–9
competitivity 51
Connecting the Dots for Digital Transformation in Agri-food 72
contractual governance 80
convergence research and innovation (CI): in agri-food context 113–14; behavioral analytics and 116–18; digitization and 115–16; framework *114*; generally 112–13, 122–3; human behavior and 114–15; modularity in 116; multi-stakeholder decision support 120–2; person-centered ecosystems 118–20, *120*
cooperation in business ecosystems 78–9
coopetition in business ecosystems 78–9
crop monitoring 27, 103
Cyber-Physical Systems (CPS) 26

Dartmouth Conference (1956) 101
data analysis 102
data analytics 27
DATA-FAIR 61
designers 77
DG Agri (EU) 71–2
DG Connect (EU) 71–2
digital agriculture 25
digital business strategy (DBS) 17–18
"digital center" 21

Digital Innovation Hubs (DIHs) 121, *122*
digitalization: Big Data Analytics (BDA) 15–17; Big Data (BD) 15–17; business impact of (*see* business impact of digitalization); business models (BM) and 11–12; challenges and strategies 4; collaboration and 18; concept of 3–4; defined 11–12; definitions in 3–4; digital business strategy (DBS) 17–18; dynamics of business ecosystems 4–6; e-market and 20; environmental issues and 26; generally 1–3, 20–1; health and safety issues and 26, 111–12; information transfer and 19; innovation and, 20; in internal processes 18; knowledge management and 19; literature review 12–20; managers, role of (*see* managers, role in digitalization); Precision Agriculture Technologies (PAT) and 25–6, 28, 31–2; services and 20; "smart" work and 19; in supply chains 19; sustainability and 6, 26, 111–12; theories of 3–4
digital synthetic ecosystems (SE) 119–20
digital technologies: adoption of 127–9; AI (*see* artificial intelligence (AI)); data analytics 27; diffusion of 127–9; digitalization and 11; drones 27; evolution of 6; expansion of 26–7; marketplaces 27; precision livestock 27; robots 27; smart greenhouses 26; smart irrigation control systems 27; software 27; soil, plants, and yield monitoring systems 27; sustainability and 6
digital transformation *see* digitalization
"digital vortex" 21
digitization: convergence research and innovation (CI) and 115–16; defined 11
Dow/Dupont 108
drones 27, 67

eAgrosuite 93
ecosystem-as-affiliation 78
ecosystem-as-structure 78
eKutir: eAgrosuite 93; farmer intervention groups 94; Farmer Portfolio Management Tool 93; generally 93, 97–8; social brokerage among rural farmers *96*, 96–7, **97**; social impact of 94; VeggieLite intervention 94–6, *95*

Elaisian: artificial intelligence (AI) and 109–10; competition analysis 108–9; generally 6, 106–7; machine learning 109–10; market analysis 107–8; predictive analytics 109–10; technological readiness 109
e-market, digitalization and 20
employees: case studies 52–3; training 53
ethics: business ecosystems and 63; ethical purchasing groups (GASes) 45–6
European Agricultural Machinery Association (CEMA) 71–2
European Commission (EC) 62, 64–5, 71–2
European Federation of ICT in Agriculture (EFITA) 61
European Federation of Trade Unions 78
European Liaison Committee for the Agricultural and Agri-Food Trade (CELCAA) 78
evaluation of technology 53, 57

Facebook 69, 72
Fall Line 107
Farmer Portfolio Management Tool 93
Finistere 107
FIspace 72
"5.0 Era" 26
FIware Foundation 72
"4.0 Era" 11, 25
4G 67
"Fourth Industrial Revolution" 11, 112, 122
Future Internet (FI) 62–3, 72
future research 127–9

Google 69, 108
Grand Challenges India 94
greenhouses: smart greenhouses 26; solar greenhouses 50
Grounded Theory (GT) methodology 38

health issues, digitalization and 26, 111–12
High Technology Integration (HTI) enterprises 39–46

ideators 77
India: agri-food in 93; Biotechnology Industry Research Assistance Council of India 94; eKutir (*see* eKutir); Grand Challenges India 94; nutrition in 111
Information and Communication Technology (ICT) 28

information transfer: in business ecosystems 63; business impact of digitalization and 28; case studies 51–3, 57; digitalization and 19; in SMEs 74
innovation: case studies 56–7; digitalization and 20; future research 127
Innovation Endeavors 107
Innovation Experiments 121
Innovation Portals 121
Innovation Services Maturity Model 121
integration of new ideas 1–2
intellectual property rights 2
intermediaries 77
internal processes, digitalization in 18
International Agri-Food Network (IAFN) 78
Internet of Food and Farm (IOF2020) 2–3, 5, 61, 64–5, 67–8, 72
Internet of Things (IoT) 26, 62, 67, 73
irrigation control systems 27

keystone players 76
Khosla 107
knowledge management, digitalization and 19

lead organization governance 82, 85
LinkedIn 72
literature review: business impact of digitalization 27–8, **29–30**; data extraction 14; digitalization 12–20; grouping 12–14; list of questions **15**; methodology of 12–15; results of 15–20
LoRa 67
Low Technology Integration (LTI) enterprises 39–46

machine learning 109–10, 119
managerial factors 2
managerial style 54
managers, role in digitalization: generally 37–8, 46; High Technology Integration (HTI) versus Low Technology Integration (LTI) enterprises 39–46; managerial behavior 42–3; managerial beliefs 39–41; managerial cognition 41–2; managerial practices 43–6; methodology of study 38–9
market analysis 107–8
marketplaces 27
mission statements 50

modularity 116
monitoring of technology 53
multi-stakeholder decision support 120–2
natural language processing 119
Network Administrative Organization (NAO) governance 81, 85, 87
networks: business ecosystems and 74–6; future research 128; informal networks 51, 55–6
neuroscience 118
niche players 76
nutrition 111

open firms 51
organizational learning 56–7
owners, role in digitalization *see* managers, role in digitalization
participant perspectives 2
PAT *see* Precision Agriculture Technologies (PAT)
path dependency 1
person-centered ecosystems 118–20, *120*
physical dominators 76
platforms 76
polar types approach 2
practices, managerial 43–6
Precision Agriculture Technologies (PAT): artificial intelligence (AI) and 103–4; case studies 50; digitalization and 25–6, 28, 31–2; generally 67, 127
precision livestock 27
predictive analytics 103, 109–10
productivity 50

Regional Clusters 121
relational governance 80
resilience 51
robots 27, 103
rural sector, challenges and strategies 4

safety issues, digitalization and 26, 111–12
Scopus 27
services, digitalization and 20
S2G 107
shared governance 81–2, 85

small and medium-sized enterprises (SMEs): business ecosystems (*see* business ecosystems); challenges and strategies 4; generally 2–3; information transfer in 74; managers, role in digitalization (*see* managers, role in digitalization); statistics 74
Smart Agriculture *see* Precision Agriculture Technologies (PAT)
SmartAgriHubs 2–3, 5, 61, 64, 72, 121
SmartAgriHubs Foundation 73
Smart Farming Technologies *see* Precision Agriculture Technologies (PAT)
smart greenhouses 26
smart irrigation control systems 27
Smart Solutions 72
smart use of information 51
"smart" work 19
SMEs *see* small and medium-sized enterprises (SMEs)
software 27
soil monitoring 27, 103
solar greenhouses 50
strategic renewal 1
supply chains: artificial intelligence (AI) and 102–4; case studies 55; digitalization in 19; supply chain management (SCM), 19, 103
sustainability: of business ecosystems 72–3; digitalization and 6, 26, 111–12; future research 128; as pillar of business model 51; United Nations and 112

technological readiness 109
transaction cost economics (TCE) 79–80
trust 81

Uber 68
United Nations, sustainability and 112

value dominators 76
VeggieLite intervention 94–6, *95*

Wageningen Digital Innovation 5
Wageningen Economic Research 61–2
Wageningen University and Research (WUR) 61–2, 64
Web of Science (database) 12, 21
Wi-Fi 67, 106